16-43

HISTORY OF THE GRINDING MACHINE

WHEEL DIMENSIONS

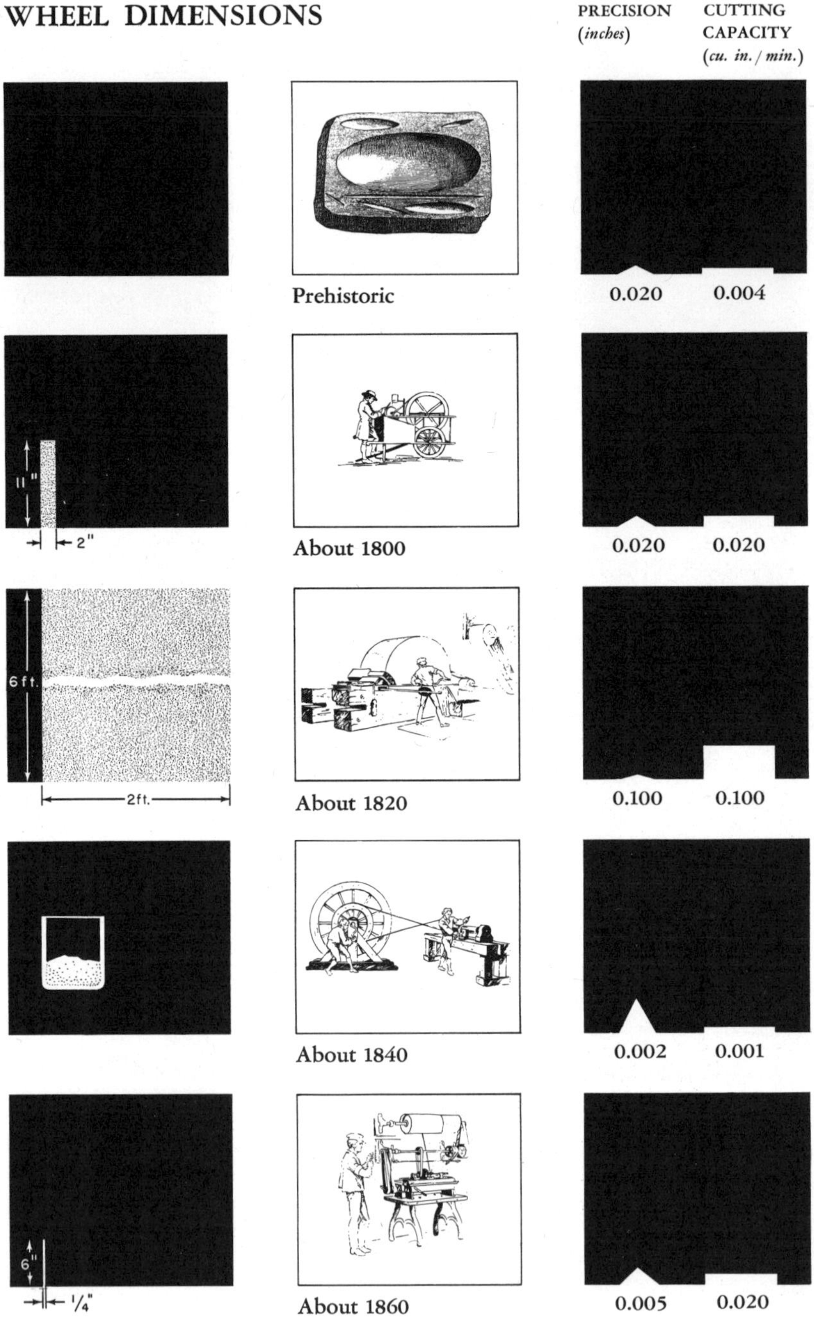

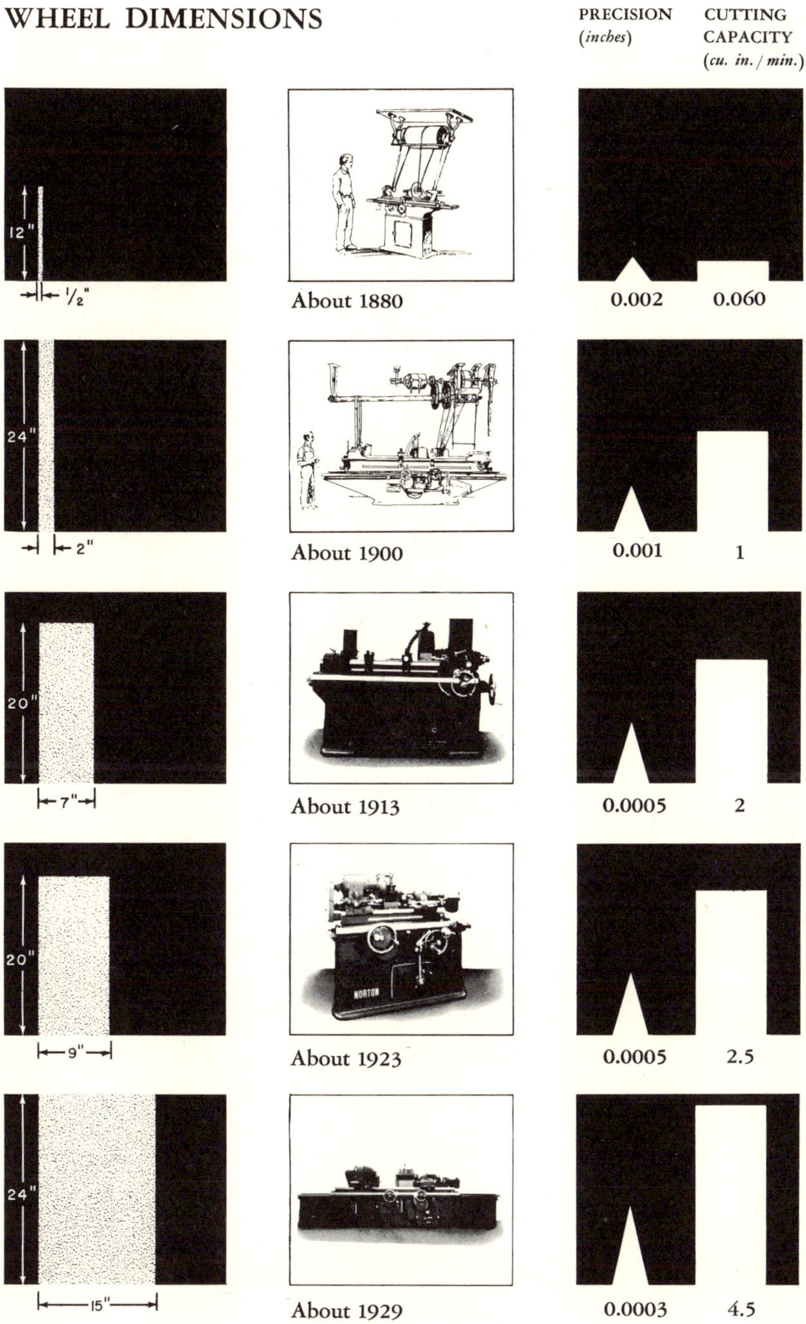

WHEEL DIMENSIONS
PRECISION (inches)
CUTTING CAPACITY (cu. in./min.)
12"
1/2"
About 1880
0.002
0.060
24"
2"
About 1900
0.001
1
20"
7"
About 1913
0.0005
2
20"
9"
NORTON
About 1923
0.0005
2.5
24"
15"
About 1929
0.0003
4.5

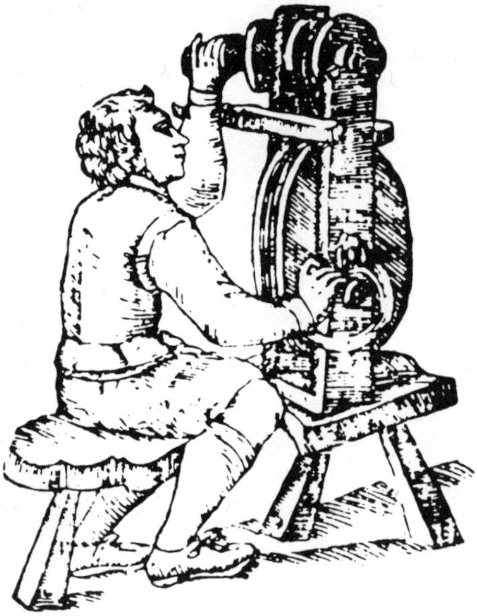

Robert S. Woodbury

HISTORY OF THE GRINDING MACHINE

A Historical Study in Tools and Precision Production

THE M.I.T. PRESS
Massachusetts Institute of Technology
Cambridge, Massachusetts

Second Printing April, 1964

Library of Congress Catalog Card Number: 64-23020

Design by
Jacqueline Shepard Casey

To
HOWARD RUSSELL BARTLETT
for
many quiet kindnesses
over the years

The publication of this study
is made possible
by generous grants from
The Cincinnati Milling Machine Company,
Cincinnati, Ohio,
and the
Norton Company,
Worcester, Massachusetts

TABLE OF CONTENTS

TABLE OF CONTENTS *continued*

PREFACE

This monograph is the second in a series intended to provide a sound foundation for writing a History of Tools. It continues the basic aims indicated in the Preface to my *History of the Gear-Cutting Machine,* and will be followed by a monograph on the History of Shop Precision of Measurement. In each of these monographs it has been possible, without detracting from the principal topic, to indicate the many ramifications of the History of Technology in even as restricted a field as the History of Tools. In my *History of the Milling Machine* I emphasize the purely technical development of the tool itself; in the *History of the Gear-Cutting Machine* I tried to show the relationship between a rather complex theory and the machines designed to embody it in metal. In the present monograph I have attempted to indicate the historical influence of a given tool on the industrial production which it makes technically possible. In the study of Shop Precision of Measurement it is possible to analyze the interaction of precise gages and machine tools with the principle of interchangeable parts, which lies at the very core of the most important of our methods of mass production. In later monographs the development of high-production automatic machine tools and the results of science applied to metal cutting will be related to our industrial production techniques.

After I had already completed preliminary studies on the early history of the grinding machine, my attention was kindly drawn by Mr. H. T. Pledge, Keeper of the Library of the Science Museum, London, to the Braunschweig dissertation of Alf Schroeder, *Entwicklung der Schleiftechnik bis zur Mitte des 19. Jahrhunderts,* Verlag Petzold-Druck, Hoya-Weser, 1931. This is one of the very few scholarly monographic studies on the History of Tools, and corrects, completes, and supersedes the accounts of Franz Feldhaus in his article "Schleifstein" in his *Die Technik der Vorzeit, der Geschichtlichen Zeit und der Naturvölker,* Leipzig, 1914, and in his *Die Geschichte der Schleifmittel,* Hanover-Hainholz, 1919. From classical antiquity onward Schroeder presents the whole subject of grinding most thoroughly. I therefore felt fully justified in relying on his work for a few of the principal facts in the development of the grinding machine up to the middle of the 19th century.

Because Schroeder's study may not be easily available to many readers, I have included some of his work in my first two chapters, together with an analysis of my own, in order to present a fairly complete account of the development of the grinding machine.

My special thanks for assistance in obtaining access to source materials are due to Dr. Philip W. Bishop of the Smithsonian Institution, to Mr. A. William Meyer of Brown & Sharpe, to Mr. E. T. Larson, Mr. A. Belden and Mr. H. W. Eaton of the Norton Company, and to Mr. Sam Redrow, Jr., and Mr. Mario Martellotti of the Cincinnati Milling Machine Co.

ROBERT S. WOODBURY

Cambridge, Massachusetts
August 1, 1958

INTRODUCTION

The grinding process of shaping materials to fit men's needs is one of the earliest we know. Relatively soon after man had learned, first to break, then to chip, and then to flake stones to get his first cutting tools, he learned also to grind stone and other materials to form tools and other useful objects. In fact, from the Neolithic Period (15,000 to 5000 B.C.) onward man has used the abrasive properties of a number of naturally occurring substances, such as sandstone, emery, and jewels, to cut and form other very hard materials.

The invention of the grinding method of making tools enabled early man to make stone hand tools far more sophisticated and more efficient than he had been able to produce by the cruder methods of fracture. Since in grinding he was less restricted by the crystalline structure of the stone, he could make his tool of a size and shape better adapted to his purpose and in making his tools have fewer failures caused by incorrect estimate of where and how to strike or apply pressure to his raw material. In addition, grinding allowed early man to utilize for his purposes many substances otherwise too hard or without the convenient structure of flint, since he could grind anything which was softer than his abrasive material. Grinding also permitted him to drill a hole through stone and to produce grooves in the stone at desired places; he could then fasten a handle to his tool in a far more satisfactory way. An adequate handle, of course, increased the efficiency of the tool immensely. So Neolithic man could make an excellent stone hammer or axe.

But far more important was the fact that the grinding process permitted prehistoric man to sharpen his tools to a cutting edge much superior to that of the chip or flake tools. And what was more, this edge was easily renewed. By the use of grinding techniques early man could, then, make excellent stone axes, knives, and other cutting tools out of nearly any hard material available to him (Fig. 1). It is therefore clear that not only was grinding a very early process in the history of tools, but it had at the beginning

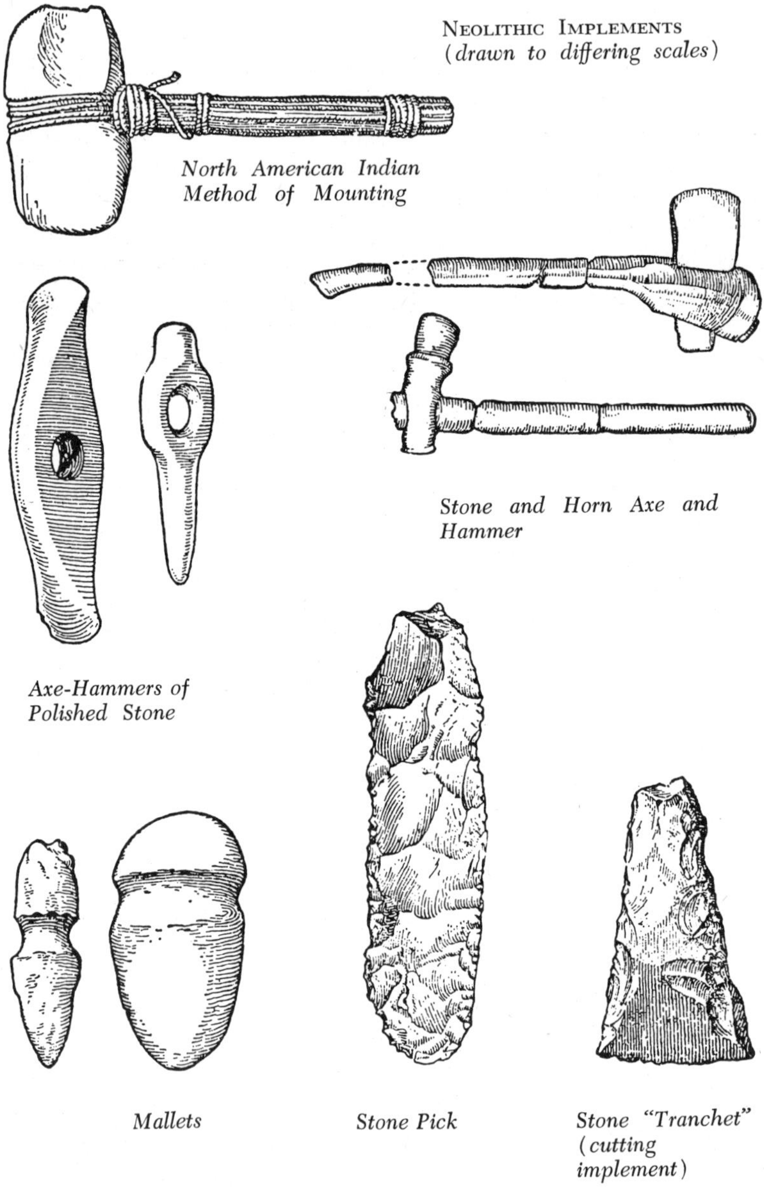

FIG. 1 NEOLITHIC TOOLS, CHIPPED, FLAKED, AND GROUND
(*Wells, Outline of History.*) *Garden City Publishing Co.*

of the history of civilization an importance not to be underestimated—probably even greater than the significance it has come to have in production in the twentieth century.[1]

Flint Arrowheads

We have seen that grinding enabled man very early in his history to utilize the *hardest substances* for his own purposes and to *sharpen* his tools and implements. Throughout the history of grinding these have been two of its basic uses. The use of grinding to give a *fine finish* to a surface for the sake of appearance may have been in the mind of Neolithic man; it certainly was an element in the grinding of stone done by the dawn of civilization (circa 3500 B.C.). So we shall not be surprised to find that the grinding machine, one of the most recent of the machine tools to come into industry, gives us even now these same characteristics in metal-cutting. But it gives us still more—*precision* and *rapid production* in working all kinds of materials.

In this monograph, however, we are primarily interested in grinding as a method of cutting metals. We shall therefore ignore here its very important uses in woodworking[2] and in the more specialized working of stone, glass, and gems. These also have a long and interesting history,[3] but one which is of less significance than the application of grinding to working hard metals. With the decline of the Stone Age into the Bronze Age (circa 5000 B.C.) grinding probably became less important in the history of tools, for the copper and bronze tools of this period could be easily formed by the new technique of casting, and to sharpen their cutting edges by hot or cold forging was simple and adequate for most purposes. Forging also actually hardened the edge to some degree.[4] With the beginning of the Iron Age (circa

Flint Knife

1. This early development will be described in detail in a later work on the History of Hand Tools.
2. The author's projected History of Woodworking Tools will include this use of abrasives.
3. For the period from antiquity to mid-19th century much information on these processes can be found in Schroeder.
4. It should, however, be noted that the bronze razors of the Egyptians probably had ground edges. The production and sharpening of tools by the techniques of casting and forging will be treated in a later work on the History of Metal-Forming Tools.

1100 B.C.) grinding again became important for tools. Available furnace temperatures did not permit casting in iron until the late Middle Ages, and although iron tools could be formed by forging, they had to be sharpened by grinding.[5] Some hardening of steel for weapons and armor was known in antiquity, and the process had developed by the fifteenth century to a rather high technical level, but hardened steel parts had not appeared in machinery even by the days of Agricola. Leonardo da Vinci's extensive series of drawings for making needles by machinery only permitted them to be sharpened and polished by grinding. The grinding of metals was actually, until the end of the eighteenth century, largely a *sharpening* or a *polishing process.*

It remained, however, a simple hand process until long after the grinding of grain had been mechanized. Mechanization of metal grinding required two basic elements: *a*) a rotating circular grinding wheel, eventually driven by power,[6] and *b*) some means of holding the work against the wheel and guiding their relative motions mechanically. Although the first steps in both these directions had been taken by the middle of the fifteenth century, the first grinding machine that we could properly call a *machine tool* did not appear until about 1830.

The use of grinding as a means of obtaining *precision* does not appear in complete form until after the middle of the 19th century with the work of Joseph R. Brown. In this characteristic the grinding machine is unique among machine tools, for it is the only one capable of producing work of the same precision as its own parts. It is also exceptional in that it is one of the few machines originally de-

5. The later use of grinding as a means of sharpening tools will be included in a monograph on The Metal-Cutting Process and Its Tools.
6. A rotating, circular stone, with its axis vertical, used for grinding grain had appeared in the Roman quern, but here the grinding was done on the face of the stone. When the principle of this device was later used for grinding metals, the stone was more commonly used with its axis horizontal and the grinding was done on the stone's edge. Such a transition logically would be expected to occur very quickly; the fact that this one took about 1000 years will not, however, surprise the historian of early technology.

veloped to improve the *quality* of its product rather than to increase the rate of production. Grinding as a means of *rapid production* had to wait for Edward Acheson and Charles Norton at the turn of the century. We shall be concerned in this monograph primarily with these later developments, which led to modern grinding as done in a machine tool capable of giving fine finish and high precision in rapid and economical production cutting of even the hardest metals.

The grinding machine is of special interest in the study of the history of types of tools. It appeared very early as the simple, hand-driven grinding wheel, and remained in only that form, although power driven, for many centuries as a machine capable only of sharpening and polishing hard substances held by hand. As soon as it is conceived as capable of high precision, it finally emerges *as a machine tool* very nearly full blown. It had passed by a sort of metamorphosis through the development of the lathe—rigidity of construction, precision, and automatic operation—while still in the form of other machine tools. In fact, the appearance of the grinding machine as a metal-cutting machine tool, as a late arrival, is rather startlingly advanced, because most of its technical problems had already been solved for other machines. All that was required was to adapt them to the peculiar requirements of the grinding machine. The really crucial problems of the grinding machine were not to provide a strong precision bed—that had been done for the lathe. Nor was it necessary to develop a head and tail stock and means of turning the work. Nor were accurate lead screws in longitudinal or cross feed new—all these, together with automatic and variable feeds, were already on the lathe and other well developed machine tools.[7] Even the swivel table for the support of the work was already in the milling machine.[8] All these elements were available to Joseph R. Brown to be incorporated in his universal grinding machine of 1868.

7. For the development of the lathe see the brilliant and thorough doctoral dissertation of Karl Wittmann, *Die Entwicklung der Drehbank,* VDI Verlag, Berlin, 1941.

8. See the author's *History of the Milling Machine,* The Technology Press, Cambridge, Mass., 1960.

The grinding machine uses a basically different cutting tool. The lathe, shaper, planer, slotter, and boring machine use single-point tools. The drill press uses a double-point, and the milling machine, broach, and bandsaw use multiple-point tools. But the grinding machine uses thousands of points simultaneously and millions of tiny points continually. This fact makes it in many ways a quite different machine tool. Yet the means by which it obtains and controls contact of its cutting tool with the work has used the methods of all these other machine tools.

The critical technical problem that had to be solved for the grinding machine was the wheel itself. The solution was finally found to be to abandon the naturally occurring abrasives and to make wheels using the improved artificial abrasives in a matrix of various bonding materials. The problem of the bonding material was attacked in several different ways by a number of men after 1870 and is still an important problem in the abrasive industry. Acheson produced the first artificial abrasive, silicon carbide, in 1891. A few years later artificial abrasives of aluminum oxide appeared. More recently the diamond and boron carbide have become important abrasives.

But for the grinding machine to become the production tool it is today, it was necessary to convince machinists that, properly designed and with proper choice of abrasive wheels and speeds, it could be something more than its French name, *machine à rectifier*, still indicates; that it could do more than give outstanding surface finish, more than give final precision, usually to hardened surfaces. It was the vision and design skill of Charles Norton which made the grinding machine into a basic production machine tool capable, not only of precision, but of rapid and economical metal cutting.

The development of the grinding machine as a production tool in industry is of special interest, for it illustrates succinctly the dangers of accounting for economic and social change without considering carefully and in some detail the technological basis which is a necessary condition of industrialization. Of the various technological advances which made our industrial society possible, that most neglected by economic and business historians, yet the one which is frequently crucial, is the machine tool. Without the grinding machine, three important American industries would not have been possible—the sewing machine, the bicycle, and the automobile.[9] The economic importance of the automobile is today all too self-evident; the social importance of all three of these inventions is well known to historians.

After the invention of a practical sewing machine by Elias Howe in 1846, and some later improvements in design, the question was how to manufacture it in quantity and of a quality to satisfy the market. The potential customer was a woman, and she intended to use her sewing machine in her home.[10] This meant that it could not resemble the cumbersome and noisy machines of all types tolerated in factories of the day. It must be light, smooth, and quiet in operation. Equally the sewing machine involved a substantial investment; it was one of the first home appliances to be "financed." It must therefore last a lifetime or more; and it must be trouble free, since no mechanic would be standing by to adjust or repair it. The only way in which all these requirements could be met was to use many hardened steel parts working with each other in close fits. It was already known that hardening of a steel part, especially if of an irregular shape, produced appreciable distortion. The only solution was, in the initial machining, to allow for the distortion, harden the part, and then grind it to the finished dimensions and quality of surface. Some of the parts also required a fine surface finish. The surface desired could be obtained by hand grinding, but the precision dimensions required a grinding *machine tool.* As we shall see, Joseph R. Brown met this demand, and light grinding machines not very different from his original universal grinding machine[11] are in use today in

9. The important application of grinding techniques to the improved manufacture and maintenance of the railroad locomotive will be considered later.

10. The sewing machine was the first piece of complex machinery to enter the home since the clock. The clock had had similar technical requirements, but most of them were met incidental to its basic need for accurate time keeping.

11. On exhibit at the Brown & Sharpe plant, Providence, R. I.

many factories making small, hardened steel parts of precision for sewing machines, small arms, and a thousand similar devices.

It is difficult for us today to realize what an important industry the bicycle gave rise to in the 1890's, for the bicycle industry has about gone in America, though it is still important in many European countries. But the bicycle craze at one time supported a significant industry and one which had important connections with the early history of both the automobile and the airplane. The bicycle, while not as complex or expensive a device as the sewing machine, had many very similar technical requirements. In addition, it had hardened ball bearings and their races to be ground to precision dimensions and to fine surfaces to eliminate friction; and the chainless bicycle required hardened and ground gears for which a specialized gear-grinding machine was developed.[12]

But grinding processes for the sewing machine and for the bicycle required only a relatively light machine tool, for the parts were small and the cuts taken were light. The demands put on the grinding machine were still only that it work hardened steel with precision and fine finish. It was Charles H. Norton and the automobile industry which brought the next great change in the grinding machine, for without it the automobile, mass produced and of a price and reliability that made it available to all, would not have been possible. Because the automobile was a far more complex device than the bicycle or the sewing machine, it had all of their technical problems and many more of its own. As soon as Henry Ford decreed that his car must be light and cheap, its wearing parts in both engine and chassis had to be made largely of hardened alloy steels. These could have been ground to the necessary precision on enlarged and specialized models of the light manufacturing grinding machines then available, but the cost would have been prohibitive. Charles Norton was to show how hardened

12. See R. S. Woodbury, *History of the Gear-Cutting Machine*, The Technology Press, Cambridge, Mass., 1958, p. 122.

steel parts of *substantial size* could be produced *rapidly* and *cheaply,* and with the necessary precision and finish—by grinding.

When the manufacture of automobiles came to be a matter of mass production, the use of large numbers of highly specialized machine tools could be economically justified, and the special needs of the automotive industry came to have a very profound effect upon the design of all machine tools, and no less upon the grinding machine. The Heald eccentric internal cylinder grinder[13] is a case in point. Both industries had therefore very important technical and economic influences upon each other. In more recent years a similar influence has been exerted by the aircraft industry, where the very high precision required and special alloys used have made grinding of even greater importance.

Grinding also played an important part in the history of automatic machine tools. The most important development was the principle of centerless grinding, in which the work is supported not on centers, but between two opposed grinding wheels. (Fig. 72a). From 1921 the Cincinnati Grinding Machine Company, developed the technique of centerless grinding[14] so that it became the basic method of producing many relatively small, very high precision parts, both hardened and otherwise, at very high production rates and at low unit cost. This method was further extended by the Landis Machine Company into what had been a field of the automatic screw machine, to produce centerless screw-thread grinders, as well as many specialized grinding machines for the automotive industry.

By all these fundamental advances the grinding machine of Joseph R. Brown became a high-speed precision tool of first importance in both light and heavy production.

The technological growth of the grinding machine is illustrated graphically in the figures on the endpapers of this book, where the changes in the machine itself and in the size and proportions of the grinding wheel used on it are

13. See p. 130.
14. See p. 151.

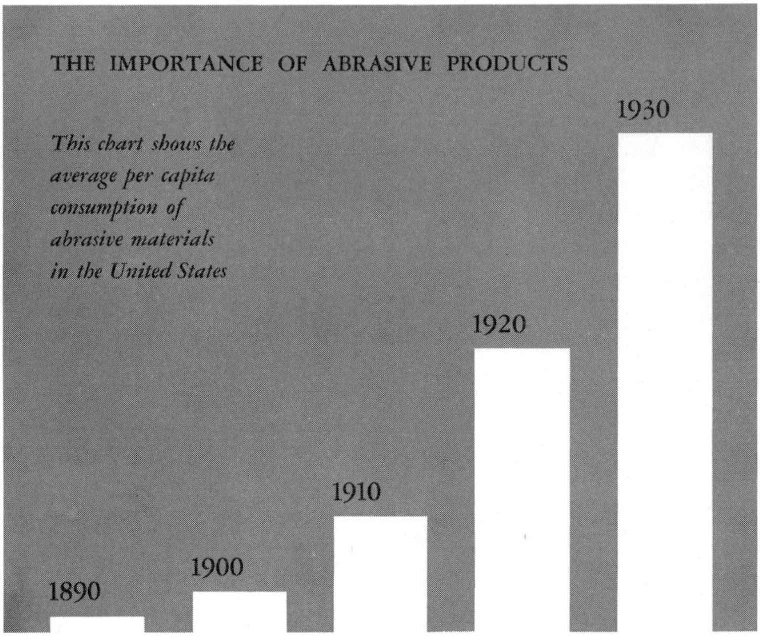

FIG. 2 IMPORTANCE OF ABRASIVES IN INDUSTRY (*Norton Company*)

coordinated with the resulting improvements in precision and cutting capacity. The tremendous rise in cutting capacity after Charles Norton's day should be noted, as should the steady increase in precision after Brown. The industrial significance of the grinding process is indicated in Figure 2, which shows the enormous increase in *per capita* use of abrasives after 1910. The sharp rise in consumption of abrasives after 1910 stems from the influence of the automobile industry; a similar sharp rise which occurred after 1930 would reflect the widespread use of grinding in the aircraft industry.

In short, in the development of the grinding machine we shall find an ancient process adapted by modern mechanization and by the results of applied science to a tool which has made possible technological changes of vast economic and social significance.

EARLY MECHANIZATION OF GRINDING

The Grinding Wheel

From the beginnings until the Middle Ages there were only two techniques of grinding, both hand methods. One was to hold the work in the hand and grind it on a fixed, usually flat, grindstone. The other was to take a smaller piece of the grindstone in the hand and rub it against the fixed work. Before a grinding machine was possible these two hand methods had to be mechanized, and this required, as mechanization of a hand process usually does, the use of a quite different principle—the replacement of a reciprocating hand motion by a rotary mechanical motion.

The first mechanization of grinding is to be found in the simple grinding wheel, then as now a tool used principally for sharpening and polishing, with the work held in the operator's hand. It does not appear, however, until the 9th century,[1] when it is shown in the Utrecht Psalter of 850 (Fig. 3). This device provides a rotating grindstone turned by a crank. We have therefore, at least this early, one of the principal elements of the grinding machine—a rotating grinding wheel mounted on a horizontal axis and cutting metal, usually on its periphery.

We should also like to be able to establish when and where the grinding machine with a treadle drive and with power drive first appeared. Schroeder gives a number of references from the 14th century, but concludes that none

1. The grinding wheel operated by a treadle and crank mechanism shown in Schroeder (from A. Rich, *Wörtenbuch der römischen Altertümer,* Paris, 1862, p. 194) as engraved on a gem of Roman times is clearly incorrectly dated.

FIG. 3 GRINDING WHEEL IN UTRECHT PSALTER, 850 A. D.
(*Library of University of Leyden*)

is in sufficient detail for us to be sure just what sort of grinding wheel is involved. The first clear evidence we have for the treadle grinding wheel is to be found in a copper engraving (Fig. 4) made about 1485 by Israhel van Meckenem.[2] This late date for the first use of the familiar treadle grinding wheel is consistent with the appearance, about this same time, of treadle drive in other devices, such as the spinning wheel. After the 15th century, illustrations of treadle grinding wheels, frequently portable, become common (See end papers of book).

We know that large querns were turned by animals in Roman times, and perhaps by water power as described in Vitruvius. The water-powered grinding mill for grain is known as early as the 14th century. Possibly grinding wheels for metal were turned by animal or water power at this same time, but available records are too limited in their descriptions for us to be certain.[3] The fact that the early powered grinding wheels that we do know of were used principally in the manufacture of armor would suggest a date at least a century later as more probable. Leonardo

2. Shown in Schroeder from *Kupferstichsammlung in Wien,* Vol. 222.
3. See Schroeder, pp. 27-29.

FIG. 4 GRINDING WHEEL WITH TREADLE, 1485 (*Schroeder*)

da Vinci's grinding machines of about 1500 are in some cases clearly power driven (Fig. 7), but we do not know whether they were sketches of actual practice, suggested improvements on machines of his day, or studies of his proposed inventions. We have to wait for a power-driven grinding wheel which we can date with certainty until the Nürnberg painting of 1568 showing an armor polisher working at two grinding wheels clearly driven by a water wheel (Fig. 5). By the next generation we have many drawings of power-driven grinding wheels in Zonca (1607), Strada (1618), and Böckler (1661).

The First Grinding Machine

In all the grinding devices we have examined thus far the work is held in the operator's hands. For a grinding *machine* we must provide mechanical means of holding the work and of guiding its motion relative to the grinding wheel. For the

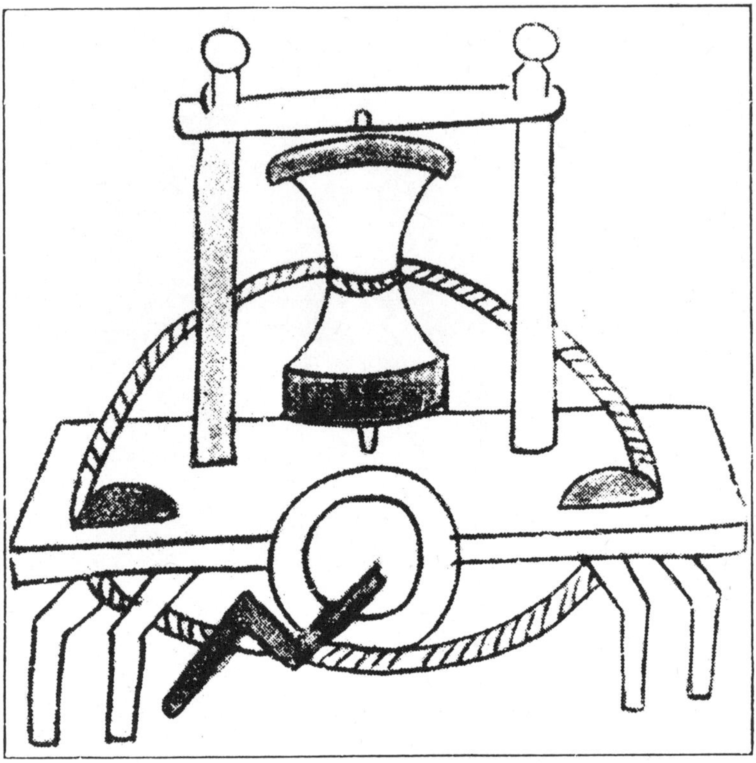

Fig. 6 Grinding Machine of About 1430 (*Schroeder*)

beginnings of this development we must go back to the early 15th century.[4]

Figure 6 reproduces a drawing from the manuscript[5] of an unknown German military engineer of the time of the Hussite War (circa 1430). The accompanying text says only:

4. After careful consideration of the evidence for a primitive grinding machine and for a rotary grinding wheel in prehistoric times given in Wibel, Müller, and Forrer, and a thorough search for evidence of these two devices in antiquity, Schroeder concludes that there is no basis on which we may say that either of them was known prior to the late Middle Ages. He apparently did not know of the evidence for a rotating grinding wheel shown in the Utrecht Psalter (Fig. 3).

5. Shown in Schroeder from Münchener Staatsbibliothek, Codex lat., Nr. 197, Folio 23v.

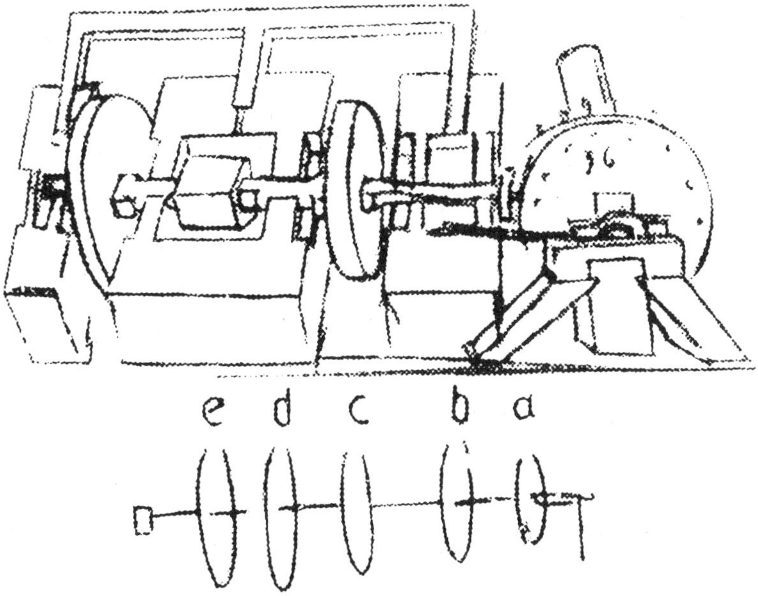

FIG. 7 LEONARDO'S POWER-DRIVEN EXTERNAL GRINDING AND POLISHING MACHINE, 1500 (*Codex Atlanticus*)

"This is a polishing mill as used by superior masters in Venice for polishing stones of all kinds. It has three wheels, the first of lead, the second of tin, and the third of copper." In interpreting this drawing one must remember that engineers of this time were interested only to show how the device was constructed and how it worked, not to represent an actual machine, much less to give a working drawing. Actual proportions and precise relative positions cannot then be assumed to be indicated. It is, however, clear that the "mill" consisted of a table on which were mounted two uprights supporting a cross beam which provided the upper bearing for the grinding wheel. A lower bearing was fitted in the surface of the table, and provision was made for conveniently changing from one grinding wheel to another. The grinding wheel was driven by a cord led over two guide pulleys to a crank-operated driving pulley.

Of course this device meets only partially our requirements for a grinding machine: a) it does have a rotating grinding wheel (in this case using emery carried by the soft metal of the wheel) and it does cut on its outer circumference, b) the work is to some degree mechanically supported, since the wheel probably ran nearly flush with the surface of the table and the work could rest on it. But it clearly does not have mechanically controlled motion of the work relative to the grinding wheel. It is only a bare beginning. Yet it is the first evidence of anything that we could call a grinding machine. For a more fully developed grinding machine we must look to Leonardo da Vinci two generations later.

Leonardo da Vinci's Grinding Machines

As in so many other important developments in the history of technology, we find in the drawings of Leonardo da Vinci fundamental advances made in the development of the grinding machines. In a series of brilliant drawings found principally in the Codex Atlanticus, he shows many machines designed for various grinding purposes. Of these, three are of special significance.

In Folio 7Rb is a sketch of a large mill for *external grinding* (Fig. 7). While the work is intended to be held in the workman's hand, it is the first reliable evidence we have of a power-driven grinding wheel, even though, with Leonardo's usual feeling for generalized mechanical concepts, he does not show the actual source of power. But he does indicate substantial construction and bearings, as well as the details of the trundle to engage with the crown gear drive. He also describes a number of types of grinding wheels, all to be used with emery—"*a*) made of walnut wood, covered with strips of thick leather on the cutting surface and with tallow and emery put on it, *b*) made of willow wood, put together in a star-like shape. Tallow and emery are put on the cutting surface. *d*), *e*), and *f*) are made from walnut wood and used with oil and emery. Always put the emery on your work." Details of wheels *a*), *b*), and *d*) are also given in smaller sketches. Leonardo's imaginative mind was already exploring different possibilities for the powdered-emery grinding wheel.

FIG. 8 LEONARDO'S DISK GRINDING MACHINES, 1500 (*Codex Atlanticus*)

Two simpler grinding machines of interest are shown in Folio 320Rb and Folio 380Vb (Fig. 8). This is the earliest evidence we have for the *disk* type of grinding machine, in which the grinding is done on the face of the wheel rather than its periphery. Leonardo shows two types, vertical and horizontal, with different kinds of simple hand drive. But of much greater significance is that in both these machines the work is supported and guided relative to the wheel *mechanically*. This device is, then, the *first true grinding machine.*[6]

6. Even though Leonardo describes it as designed to grind the edges of sheets of glass, it could equally well be used to grind wood or metal.

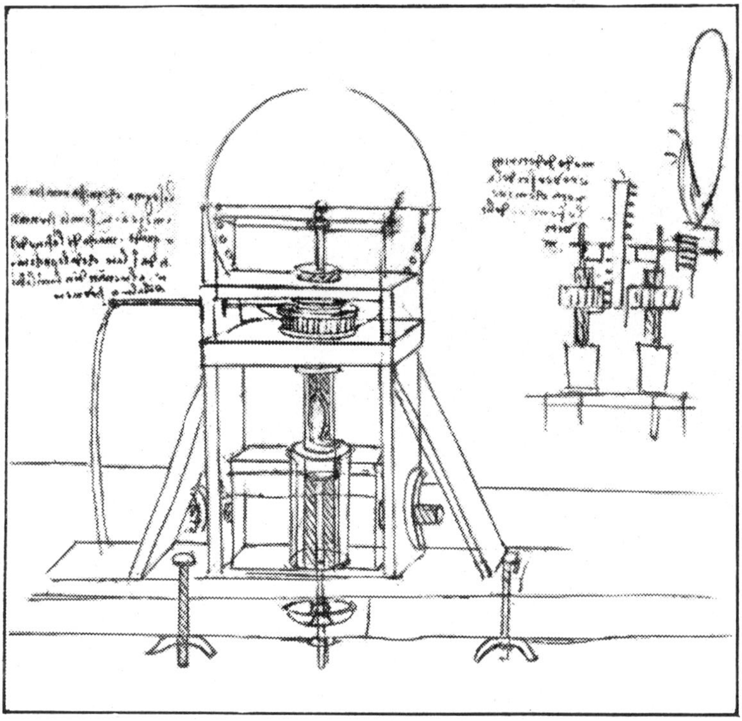

Fig. 9 Leonardo's Internal Grinding Machine, 1500 with Automatic Motion (*Codex Atlanticus*)

Leonardo also designed a machine for *internal grinding*. It is shown in Folio 291Ra (Fig. 9), and has many interesting features. The cylinder to be ground is clamped in two jaws[7] in the lower part of the machine. It should be noted that since both jaws are controlled by separate screws Leonardo could center his work accurately under the grinding shaft. This feature, combined with the solid and braced construction of the frame, allows us to claim that Leonardo had at least the germ of the concept of a *precision* grinding machine.

7. This is the first use of the screw-type vise, here in the special use of what the modern machinist would call a "fixture." The important development of means of holding the work in fixtures and guiding the cutting tool in jigs will be treated in a later monograph on the History of Jigs, Fixtures, Arbors, and Chucks.

This machine has a wooden cylindrical grinding wheel just large enough to fill the hole in the work completely, and there are grooves cut in the face of the grinding wheel to allow the oil and emery mixture to reach all grinding surfaces. The grinding wheel is secured on a driving shaft which can slide in two bearings in the upper frame of the machine. Just above the grinding wheel the squared section of the iron shaft carries a threaded piece which engages a nut in the upper part of the machine. Rotation of this nut will therefore cause an up-and-down motion of the grinding wheel in addition to its separate rotary motion in the cylinder. The nut is also carried on a trundle gear which engages an intermittent crown gear working from the main drive. The rotation of the crown gear will thus at intervals lift the grinding wheel. In the intervals when the crown gear has no teeth engaged with the nut trundle, the grinding wheel is returned to its original position by a spring-loaded cord wrapped around a pulley secured to the nut trundle. Clearly here we have a device which not only qualifies as a grinding machine, but also has *automatic motion of the grinding wheel relative to the work.*

A rough sketch at the right indicates that, as in so many of Leonardo's other engineering designs, this automatic grinding machine was intended to be *used in production in a series of machines operated by central source of power.* That Leonardo had specifically in mind the concept of a series of grinding machines in production operation is indicated in two sets of sketches he made, one for machines for grinding and polishing burning mirrors, and another for machines for grinding and polishing needles.

The series of machines for grinding the burning mirrors utilized the swinging pendulum device so often seen in other applications after Leonardo, although his machine for making a concave parabolic mirror used an ordinary grinding wheel in a most ingenious way. One machine of this series is of greater interest, for in it Leonardo used multiple grinding rollers whose axes were held parallel to each other on a curve appropriate for the desired burning

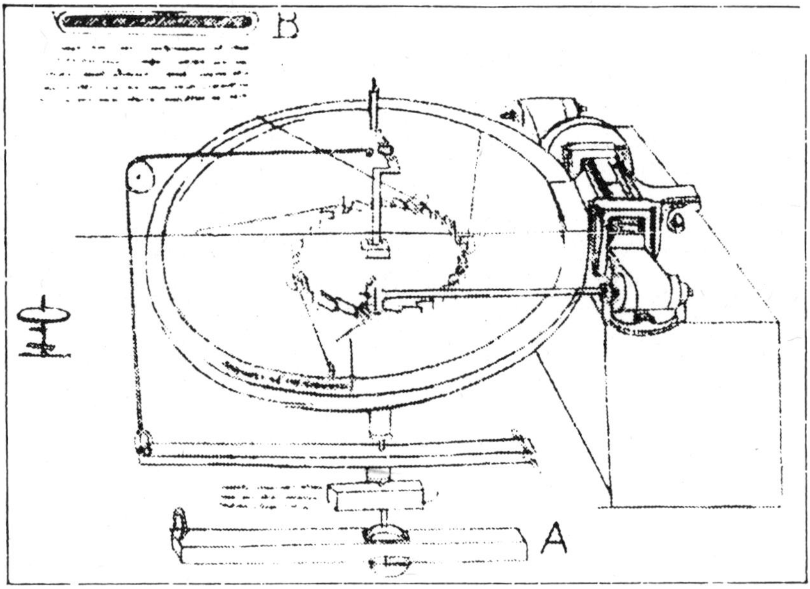

FIG. 10 LEONARDO'S BELT GRINDING MACHINE, 1500 (*Codex Atlanticus*)

mirror.[8] This is an interesting anticipation of *form grinding using multiple grinding wheels.*

The series of machines for making needles was Leonardo's pride and joy, for he expected it to make him rich. But Leonardo seems to have been as little acquainted with the economic effects to be expected from his inventions as are most engineers today. At any rate, in *Codex Atlanticus,* Folios 25, 318, and 341, he shows a series of machines intended for making needles. Their principal technical interest for us is the clear use of grinding by means of an *abrasive band* (Fig. 10). But more important is that Leonardo in this series of drawings has clearly in mind *mass production* of a *standardized product* by means of *specialized grinding machinery.*

We have, then, in Leonardo at least the germ of all the developments that were to follow in the grinding machine

8. *Institut de France MS.* G, Folio 83V.

down to 1800, and even of much that was to come later. But here again, as elsewhere in Leonardo's engineering work, he was followed by men of less vision and imagination, men unable to reach his level of generalized engineering concepts. We shall have to begin at a lower level to trace the slow later development of the grinding machine, until in the early 19th century it again approximated in practice the potentialities which Leonardo had foreseen nearly 350 years before.

Progress in the 16th, 17th, and 18th Centuries

Two generations after Leonardo we find a grinding mill not very different from the Leonardo machine shown in Figure 7. It appears in a copper engraving of 1575 by Johannes Stradanus in the *Berliner Kupferstichkabinett* as shown in Schroeder (Fig. 11). The only addition seems to be chutes to supply water to some of the wheels.[9] The other wheels may

FIG. 11 ARMOR GRINDING AND POLISHING SHOP, 1575 (*Schroeder*)

9. Schroeder shows (his Fig. 52) a woodcut of 1568 with the familiar can suspended over a treadle grinding wheel so as to drip water on the stone, the first evidence of this technique.

Fig. 12 French Armor Grinding and Polishing Shop, 1765 (*Schroeder*)

have had their peripheral surface covered with leather, as in Leonardo's grinding mill. Although there are descriptions and illustrations of grinding mills in Cardano (1557), Besson (1578), Zonca (1607), Strada (1618), and Böckler (1661), they represent no advance on the Stradanus mill of 1575. Even the Balkhauser Kotten of 1683 was only a group of similar mills under one roof and using a single source of power, to form a grinding factory. While this may be an economic advance, it is of little technical interest. In fact, by the end of the 18th century no substantial advance in machinery for general grinding had been made. A comparison of Figures 7, 11, and 12 will show how little progress was made in general grinding in 300 years. We shall have to account for this failure later in this monograph.

Two specialized grinding techniques which did make practical progress in the 17th century have a significant requirement in common: both had to produce precision grinding. And both techniques appear about the same time, the second half of the 17th century. But neither ground metals.

After the invention of the telescope and the microscope there were by the second half of the 17th century a number of amateur scientists who wanted these instruments. The nature of optical instruments required greater precision than was possible with ordinary hand grinding of lenses. Progress beyond Leonardo's concept of precision grinding, which can only be obtained in a grinding machine, then led to the *specialized grinding machines* such as the lens-grinding machine of Divini of 1660 (Fig. 13). Similar devices are found at the beginning of the 19th century designed by Edwards, Unkel, Stewart, and Legey.

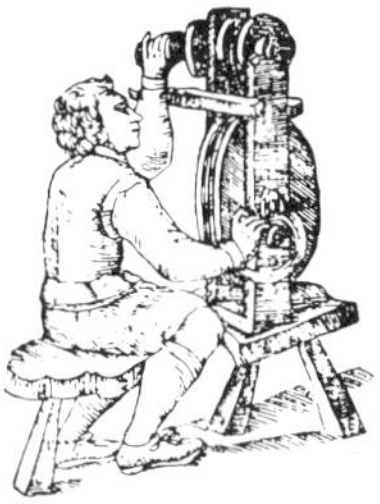

Fig. 13 Lens-Grinding Machine, 1660 (*Schroeder*)

The other technique was the grinding of precious stones as developed in Holland. Schroeder shows a copper engraving of a special grinding machine for this purpose dating from 1695. An iron disk is rotated in a horizontal plane and carries diamond dust as the abrasive. In order to grind the exact planes desired in a jewel, the stone is cemented in a fixture which is screwed tightly and held by lead weights against the rapidly revolving disk. Further progress in this specialized grinding machine is evident in a copper engraving of 1752 which shows one for three operators, as well as details of the fixture for holding the gem (Fig. 14). Simpler

Fig. 14 Grinding Machine for Gems, 1752 (*Schroeder*)

FIG. 15 NEEDLE GRINDING SHOP OF 1765, WITH DUST GUARDS (*Schroeder*)

diamond-grinding machines are shown in Diderot's encyclopedia of 1763.

Both the optical grinder and the diamond grinder are rather specialized, but they are grinding machines, and they do indicate that by the end of the 17th century at least some progress had been made in actual practice.

In the 18th century there were a number of minor improvements in grinding. There was some recognition of various grades and types of abrasives, and even some scientific study of their nature and properties. Much wider use of hand grinding arose in the expansion of the cutlery industry in England and Germany from the increased use of steel and from Hinchcliffe's invention in 1760 of the "black polish." The technique was also applied to the manufacture of all kinds of steel ornamental devices admired because of their high polish, but this industrial growth involved no important mechanical improvements. We also find in this period the beginning of protection of the workmen from effects of the particles of steel and abrasives which fly about a grinding shop. Figure 15 shows a needle grinding shop of 1765 in which boxes with glass shields help to protect the workmen's lungs from the grinding particles. This hazard was especially bad in grinding needles, since they had to be ground dry in order to avoid rust. In 1821 a prize was offered by the *Société d'Encouragement pour l'Industrie nationale* for the establishment of a needle factory in which adequate protection from the grinding dust would be provided. In 1821 Cowen had a machine for pointing the cards for a carding machine, in which the grinding dust was absorbed by an exhauster. A "magnetic" dust collector consisting of a series of cloth curtains around the grinding wheel and a set of steel magnets near the workman's mouth was invented by Abraham in 1823. By 1824 the Sheffield grinding machine with an effective dust collector was available. However, the notions of industrial hygiene of the early 19th century warranted in most grinding mills only a cloth wrapped around the grinder's face for protection. Interest in a solution to this problem continued, however, until it was eliminated by the almost universal use of oil or water on the grinding wheel for other reasons.

In the late 18th century there was a revival of two earlier grinding techniques. Burrows, Pajot-Deschaines, and James Watt all constructed machines to grind mirrors. But of greater interest, since it involved internal grinding on an industrial scale, is the boring and grinding machine of 1780 shown in Figure 16. This machine was used to bore guns and then to correct the result to more exact caliber and finish by grinding with emery on a lead piston. Here we see in practice a grinding machine used, not because the material was too hard for other means, but because greater precision and a better product resulted.

From the late 18th century onward, we can detect the first beginnings of an industrial need for a grinding machine, foreseen so much earlier by the genius of Leonardo da Vinci. But we must ask, why was this demand so long in coming? Why was hand grinding adequate for manufacturing needs for so long?

The answer in two words is—the Industrial Revolution, and all that the term means from the standpoint of *technological* advance: the enormous increase in the use of machinery of all kinds, more and more frequently made of iron and steel and with a complexity which required precision manufacture of its parts; the manufacture of iron and even steel for many purposes, which required plate and rolling mills, many of whose parts had to be accurate and yet capable of withstanding high stresses as well as high temperatures, and therefore must be hardened; the increasing use of power-driven machinery which increased the speed and size of other machinery so that both the steam engine and the machines it drove required greater precision of construction. All this gave rise to the Age of the Machine Tool: Wilkinson's boring engine, Maudslay's lathe, Roberts' planer, and Nasmyth's drill press. Two generations, from 1775 to 1840, saw Steam, Iron, and the Machine Tool produce the most profound change of all time—the Industrial Revolution. The great day of the grinding machine was to come a little later, but it had a part to play even in this early drama.

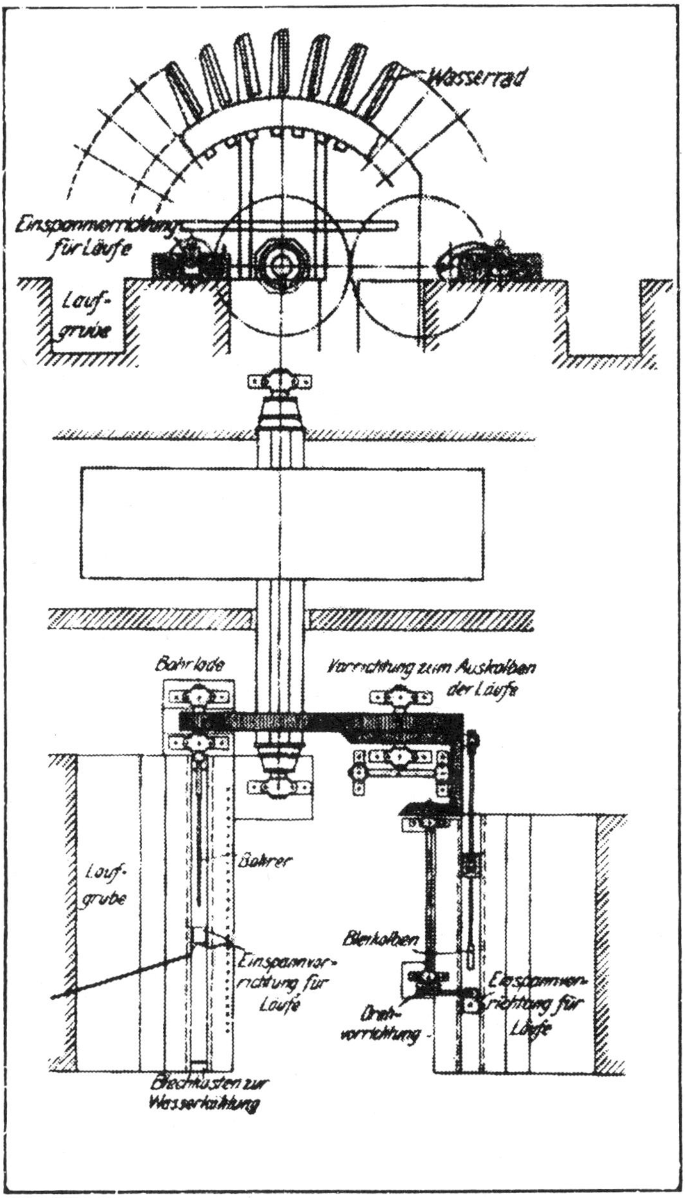

FIG. 16 BORING AND GRINDING MACHINE OF 1780 (*Schroeder*)

FIG. 17 GRINDING RIFLE BARRELS, SPRINGFIELD ARMORY, ABOUT 1818
(*American Machinist*)

THE BEGINNINGS OF THE INDUSTRIAL GRINDING MACHINE

1830 to 1860

The Armories

The need for grinding techniques in the production of guns, which had been recognized as early as 1780 (Fig. 16), was accelerated by the demands for munitions of all sorts for the Napoleonic wars and for the increased standing armies which followed. Armories sprang up also in the United States, now cut off from the supplies of England and France. As is so often the case, wartime demands forced new methods of manufacture on the older craftsmen. Perhaps the outstanding example was the method of mass manufacture by means of interchangeable parts undertaken by Blanc in France, and by Eli Whitney and others in the United States, but similar improvements in technique were to be found in all the advanced countries.

The earliest use of "draw grinding" of gun barrels is described in a letter of Thomas Blanchard dated November 28, 1818, accepting a commission to make one like that shown in Figure 17, for use at the Springfield Armory.[1] Here a very large and thick sandstone wheel was driven at 400 revolutions per minute by belting. The wheel had a thickness greater than the length of the rifle barrels. By means of a crank the workman could rotate the barrel which he forced up against the wheel by a lever worked by his body. An iron grating on the floor prevented him from slipping while exerting this pressure. This effort to do fine grinding with natural sandstones proved rather difficult because of the size and lack of precise balance of these large wheels.

By 1829 Bianchini, Director of the Armory in Vienna, described a special machine for grinding arms (Fig. 18) on a dry sandstone wheel revolving 100 to 130 revolutions per

1. *American Machinist,* Aug. 9, 1917, p. 251.

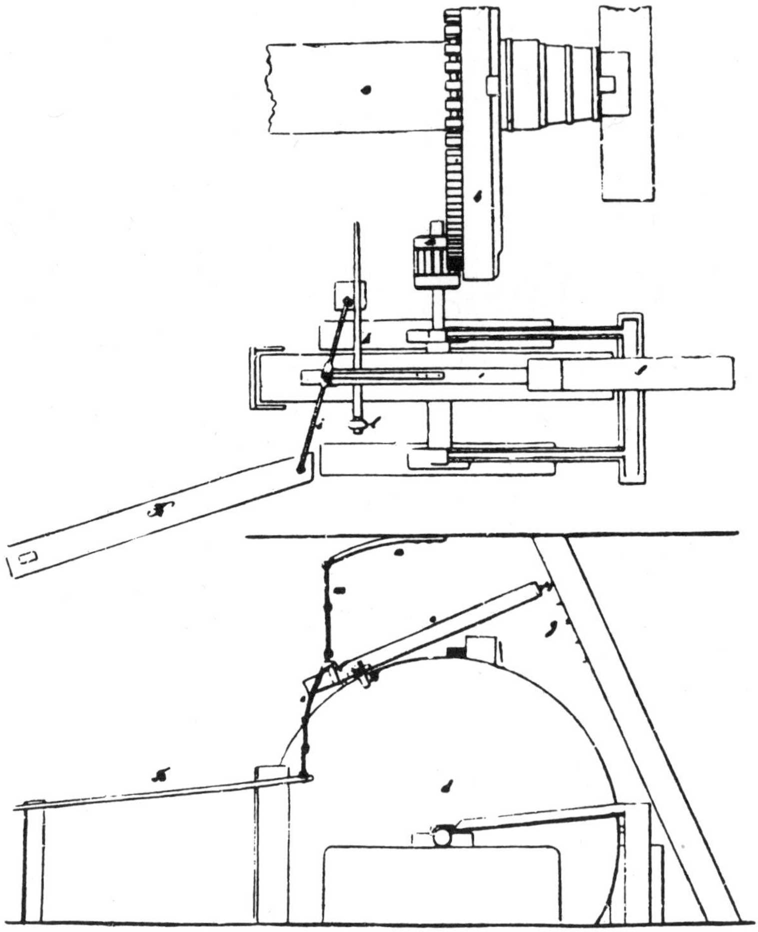

Fig. 18 Machine for Grinding Arms, 1829 (*Schroeder*)

minute. A treadle drive for rotating the barrel of the gun as it was ground was provided. By this device a workman could grind 24 to 30 barrels a day.

In Prechtl's *Technologische Encyklopedie* for 1838 is a similar machine, described as having a grindstone 7 to 12 feet in diameter and 10 to 13 inches thick, for grinding gun barrels. These very large grindstones raised for the first time, because of their great weight, the question of the support of the grinding wheel on its axis. Later, as grinding wheels went to much higher speeds, this question became of great importance in order to prevent unequal stresses in the wheel from causing it to break under the high centrifugal forces.

Admittedly these grinding machines are all crude compared to the machine of 1780, but they do at least indicate that grinding of metal was fast becoming an industrial process, other than just polishing or sharpening. In the armories it continued to improve, for we find highly developed machines for making sabers and for grinding gun barrels by 1869.

Early American Grinders

A number of American mechanics were also interested in grinding, but for more peaceful purposes. There is a tradition that as early as 1773 the hardened and polished verge wheels made in the Connecticut clock factories had their contact surfaces ground by using wheels of wood covered with leather and charged with glue and emery. These machines were of heavy oak frames, with the grinding wheel mounted on an iron bushing having a taper which fitted the taper in the bearing. By the 1840's these machines were made with iron frames.[2]

Grinding also played an important part in the early manufacture of textile machinery in America. David Wilkinson of Providence, Rhode Island, wrote "About this time [1820] . . . to make a yarn on a jenny, I made a small machine to grind with which had a roller of wood to roll on the stone, which turned the spindle against the stone, and so ground the spindles perfectly."[3] We wish that Wilkinson had told us more about his device; it could even be the first centerless grinder. But at least it is clear that he was looking for

2. Unpublished typescript of May, 1929, by Charles H. Norton in the files of the Norton Company, Worcester, Mass.

3. David Wilkinson, "Reminiscences," (written Dec. 1, 1846) in *Transactions, Rhode Island Society for the Encouragement of Domestic Industry*, 1861, p. 101. Wilkinson also invented the screw-cutting lathe with a slide rest, independently of Maudslay.

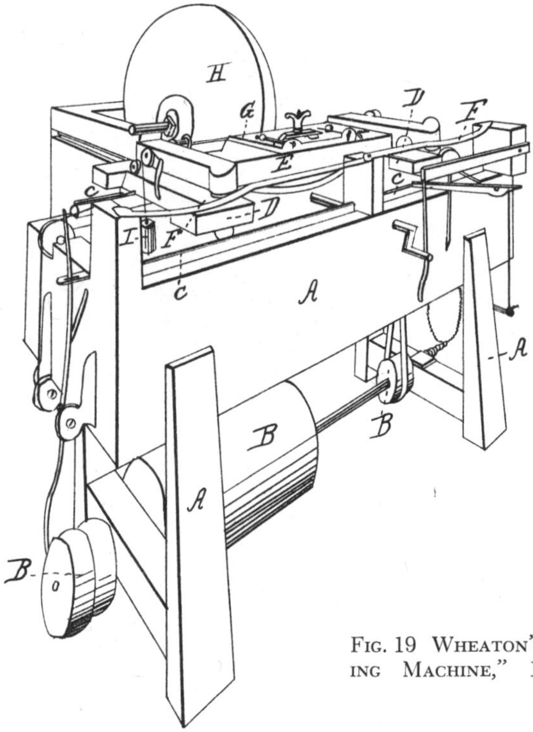

FIG. 19 WHEATON'S "ELIPTICAL GRINDING MACHINE," 1834 (*U. S. Patent*)

good surface finish, exact concentricity, and precise dimension—for it was the lack of these qualities in the spindles of his jenny which led him to try grinding.

In 1833 Jonathan Bridges described the "frequently used traverse grinder"[4] and his method of traversing the grinding wheel rather than the work. His grinding wheel was driven from a splined shaft along which it could slide. Bridges tells us that his machine was used to grind "piston shafts and card cylinders."

In the following year James Wheaton of Providence[5] in his patent dated May 28, 1834 described his "Eliptical [*sic*] Spindle Grinder" (Fig. 19) as "an Improvement on the

4 *Journal of the Franklin Institute,* 1833, p. 101.
5. Note the proximity of all these men to Slater's Mill of 1790.

Spindle Grinder in common use and which may be used for grinding various things besides spindles." This patent is so important that I quote Wheaton's description of his invention and his claims in full:

> "In the drawing hereto annexed AAA represent the frame of the machine, BBB drums on a common shaft from which motion is communicated to the carriage lathe and thing to be ground, CCC railway or slide on the top of the frame on which the carriage DD traverses. This carriage is made of iron or other hard Metal and traverses back and forward on the railway or slide, the length of the spindle or other thing to be ground so as to present every part of it in succession to the Grindstone. This motion of the carriage together with the rotary motion of the spindle or thing to be ground hereafter mentioned is derived from the drum BBB by means of both endless screws and cog wheels. Neither these parts of the machine nor their relative positions nor their operation vary from the same part of the machine, called the Spindle Grinder, now in use; on the top of the carriage is the Lathe that holds the spindle or thing to be ground. This is also in the old machine but its mode of connexion with the carriage is claimed as new, as well as the method of grinding it, and moves on the carriage from the center of the carriage or about the center, lengthwise and near the edge of it, next the grindstone extends upwardly a stud or pivot which is fast in the carriage but moves easily in a slot in the Lathe. The Lathe might be turned entirely around on this were it not for other parts of the machine, its object is to permit it to be partly turned as it moves on the carriage and at the same time to give it the same motion forward and back that the carriage has. In the old machine the Lathe has the same motion forward and back as the carriage and also a motion from side to side. The same effect may also be produced by connecting the Lathe to the carriage by means of wheels or pullies made fast to the one and moving in grooves made to the other, but this mode of connexion though new is considered inferior to the one first described.
>
> "On the top of the side of the frame most remote from the Grindstone are placed several stands extending upward as high as the lathe, attached to there is the track of the lathe FF against this track the lathe passes in its passages backward and forward on the carriage. The stud or pivot which forms its connexion with the carriage permitting it always to kick against it although it is not in a right line with the side of the lathe. The same result can be obtained by having the track under or over or on the other side of the lathe but so conveniently. This is also claimed as new. The friction of the lathe against the

track and also on the carriage is diminished by the use of friction pullies. The lathe is kept steady and pressed against the track by means of the friction straps and weights, one of which is fastened to each end of the carriage to a stud. The Strap passing up over a pulley on the lathe from which the weight is suspended; one of these is seen in the drawing and is marked II.

"On the top of the lathe and attached to it by a screw bolt passing through a slot is the spindle rest G made of lead or other soft metal. The thickness of the rest is equal to the diameter of the thing to be ground and its length depends upon the length of the thing to be ground. Against the side of this rest towards the Grindstone the Spindle or thing to be ground presses and rests. By means of two screws passing through a shoulder on the other side of this rest and playing into the stands passing through it near the other side, the Stands moving in slots in the rest, the rest itself is moved toward the grindstone as it is worn away by use. The shape, size and material of this rest are claimed as new.

"The spindle or thing to be ground is held at each end by a spring centre on which it turns, motion being communicated to it by a belt from the Drum BBB.

"H the Grindstone turns in an iron carriage on a frame extending back at right angles from the frame first described. This carriage with the stone in it can be moved on its frame to and from the spindle or thing to be ground by a screw in front of the first described frame. The friction here is diminished by the use of wheels or friction pullies under the carriage. The friction of the Grindstone and the manner in which it is moved with its carriage and turned are the same as in the old machine. The track of the lathe FF aforedescribed is made of iron or other hard metal and its shape is varied according to the shape of the spindle or thing to be ground.

"In operating a machine with this Improvement a rest is made corresponding in length and thickness with the length and diameter of the thing to be ground and is affixed to the lathe as aforedescribed. A lathe track of proper shape is affixed to the stud on the frame. The spindle or thing to be ground is then fixed on its centres in back of the rest pressing against it, Motion being communicated to the Drum. A rapid rotary motion is given the spindle or thing to be ground and the Lathe traverses back and forward on its carriage, the front part of it being kept against the lathe track by means of the weights aforedescribed and the pressure of the grindstone against the thing which is being ground, Motion being at the same time communicated to the Grindstone and the Grindstone by means of the screw being drawn and held against the spindle or thing to be ground as it passes in the lathe the requisite shape is given to the spindle or thing to be ground. The advantages resulting from this improvement are that the spindles or things ground with the same track of the Lathe are uniform in shape and size and the labour is performed with much less expense than by the ordinary method.

"What I claim as my invention is the connexion between the lathe and carriage by means of the pivot or stud placed and used as aforedescribed by which each end of the lathe is left free to move to or from the Grindstone and the whole lathe is permitted to vibrate the length of the slot in which the stud or pivot is inserted.

"The track so called which guides the lathe being of such form as to press the spindle or thing to be ground against the face of the grindstone in such manner as to produce the shape desired in the thing to be ground and also the use of weights to assist in keeping the lathe against said track and in keeping the lathe steady and the rest aforedescribed."

Some very important elements can be found in this description:

a) Wheaton and Bridges are both obviously very familiar with an earlier grinding machine many of whose characteristics can be determined from their descriptions of their own grinding machines. Perhaps Wilkinson was the inventor; certainly he and Wheaton must have known each other. But even if it was not Wilkinson, both Wheaton and Bridges tell us that the older machine had become common; surely it was in use prior to 1830.

b) The older machine already had a rotary power-driven grinding wheel, mechanical means of holding the work with at least some precision since both the ways and the carriage were of iron, and it had power feed of the work across the wheel.

c) In the older machine the work was held between and rotated on centers[6] mounted on a "lathe" on the carriage. Provision was made to feed the work in to the grinding wheel, as well as to move the grinding wheel, as it wore down, up to the work.

d) The older machine also was not restricted to grinding spindles, but was already a general-purpose machine; in

6. Note that this early these centers were spring-loaded, to take up the lengthwise expansion of the work caused by the heat of grinding.

fact it was a *complete cylindrical grinding machine*,[7] with every basic element to be found on plain cylindrical grinding machines today!

Wheaton's own improvements are even more remarkable, for he has the basic principle of the *universal grinding machine*.[8] And he has even more; his machine is fitted with a templet and holding weights such that *form grinding* can be done. It should also be noted that this thrifty and ingenious Yankee recognized that not only did his machine save expense, but it *produced a better product*.

About this same time other American mechanics were interested in surface grinding. Peter Cooper, of New York, was issued a patent March 24, 1835, for a device for grinding plain surfaces by using a tub wheel and two driving wheels to give uniform grinding on a surface held in a circular fixture. This device embodied the basic principle by which precision gage blocks are ground today.

A few years earlier a patent had been issued in 1831, to J. W. Stone[9] of Washington, D. C., for the first *surface grinding machine*. He describes his machine as suited to grinding hardened metal plate and other materials. Stone used a flat, horizontal work table which could be traversed by a screw or rack using power. Pressure was maintained on the grinding wheel by means of a spring, and "lateral vibratory motion" of the grinder spindle ensured uniform grinding of the surface. This machine was also probably the first *vertical spindle* grinding machine intended primarily for grinding metals rather than glass or stone.

Certainly the American machinists, especially those of

7. The cylindrical grinding machine grinds the outside surface of a cylindrical work piece which is held between centers, as on a lathe.
8. The principle of the universal grinding machine is described on p. 64.
9. Patent unnumbered, of Apr. 30, 1831. Stone is listed as Wm. I. Stone in *Digest of Patents, 1790-Jan. 1, 1839*, Washington, 1840 and also in the index to Dingler's *Polytechnische Journal*, Vol. 46, p. 273, but as J. W. Stone in the reproduction of his specifications in *Journal of the Franklin Institute*, Vol. VIII, 1831, p. 183, in Dingler, *loc. cit.*, and in *Repertory of Patent Inventions*, Oct. 1832, p. 214. The U. S. Patent Office has been unable to locate Stone's specifications or his drawings and believes them to have been lost in their fire.

Providence, were contributing more than their share to the new industrial technique of grinding.

Alfred Krupp

Schroeder gives a very full account of the grinding machines developed by Alfred Krupp from 1830 to 1836. In the light of the work of the American machinists we have already described and that of Whitelaw, Bodmer, and Nasmyth to be taken up in our next section, Krupp's work seems to have much less significance than Schroeder attributes to it. Certainly his claim (p. 190) that Krupp developed "the first genuine grinding machine" is hardly tenable. Schroeder attributes the shortcomings of Krupp's work to his not having available the excellent English machine tools developed in the generation before him, but the Americans had the same handicap, yet went ahead and developed their own. Of course Whitelaw, Bodmer, and Nasmyth had the best available tools of the day for their work. There are, nonetheless, features of Krupp's work which are of interest.

In his grinding machine of April 1830, Krupp was trying to develop a successful cylindrical grinding machine, but he failed to recognize the importance of iron parts to give the rigidity and accuracy unobtainable with wood, and he used wedges in the frame where an abler mechanic would surely have seen that bolts were essential. The English mechanics had seen the necessity of all this a generation earlier.

By 1831 Krupp had a better machine, which did support the work on centers. Not only was his frame inadequate and of wood, but his method of supporting the axle of his grinding wheel in wooden clamps proved unable to give the precision required for rollers for the mint. He could get them round, but not of the same diameter throughout their length. In 1832 he designed a machine to take these cylinders from his 1831 grinding machine and "rectify" them by grinding them against each other. This technique proved reasonably successful, but Krupp was called upon to make larger and larger rollers.

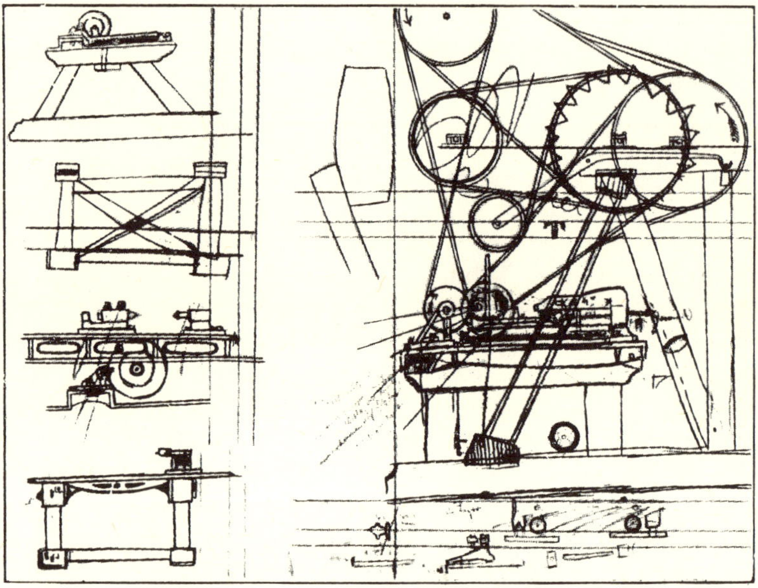

Fig. 20 Krupp's Large Grinding Machine, 1836 (*Schroeder*)

It was therefore necessary in the fall of 1836 for Krupp to design a new machine, in which all the important parts, especially the frame, were made of iron (Fig. 20). In this machine Krupp moved the slides carrying the work on large iron cylinders turned in a lathe to 4-inch diameter and 4-foot length. These cylinders were turned because he had no means of planing a flat bed of this size; in fact, Clement's "great planer" had just come into use in England. These cylinders were one of several features Krupp introduced to give his machine precision and freedom from vibration. He also took special precautions in mounting his grinding wheel, now increased from 8 inches to 20 inches in diameter. The success of this machine is indicated by the fact that it was in use for simple grinding for fifty to sixty years.

The principal interest we have in Krupp's machine is its size. It was much larger than others of its generation, the largest, in fact, until J. M. Poole's of 1868. It is also of significance in that it was clearly used in heavy industrial grinding by 1837.

In 1846 Werle, of Königsberg, had a cylindrical grinding machine, of interest because he used a disk grinding mechanism against the rotating cylinder to be ground. Werle made his grinding disk by using a large iron cup in which the sandstone was held by six set screws.

Whitelaw, Bodmer, and Nasmyth

In the 1830's we can recognize how inevitable the cylindrical grinding machine was, for it was developing not only in America and in Germany, but also in England and Scotland. James Whitelaw described in 1838 a cylindrical grinding machine[10] which he had constructed for grinding the surface of pulleys for belt transmission of power (Fig. 21). Both the grinding wheel and the work rotated, in opposite directions at the point of contact, the 42-inch wheel at 180 revolutions per minute and the pulley at 130. Convenient means were provided for feeding the work to the wheel by a hand crank, and for feed across the face of the grinding wheel by power. Whitelaw noted that this kept the surface of the grinding wheel uniform. A cover was provided to keep in the splash of the water. The machine was all of iron, and was successful in actual use. Although Whitelaw's machine is a bit specialized, it could easily be extended to general cylindrical grinding.

Of even greater interest, however, are Whitelaw's comments on the advantages of grinding over turning in a lathe, for his pulleys were of cast iron and were not ground because they had to be hardened. He points out that his machine is cheaper to make than a lathe, and that in the 18 months it has been in use it has stood up as well as any self-acting lathe. Further, it does more work: 15 pulleys of 18-inch diameter and 5-inch width in 10 hours. When the cast iron is hard, as it frequently is in thin pulleys, a lathe is good for nothing, but this makes no difference to a grinding machine. Also his machine will finish pulleys much thinner than can be turned on a lathe; therefore metal can be saved.

10. In *Transactions, Royal Scottish Society of Arts,* 1841, p. 235.

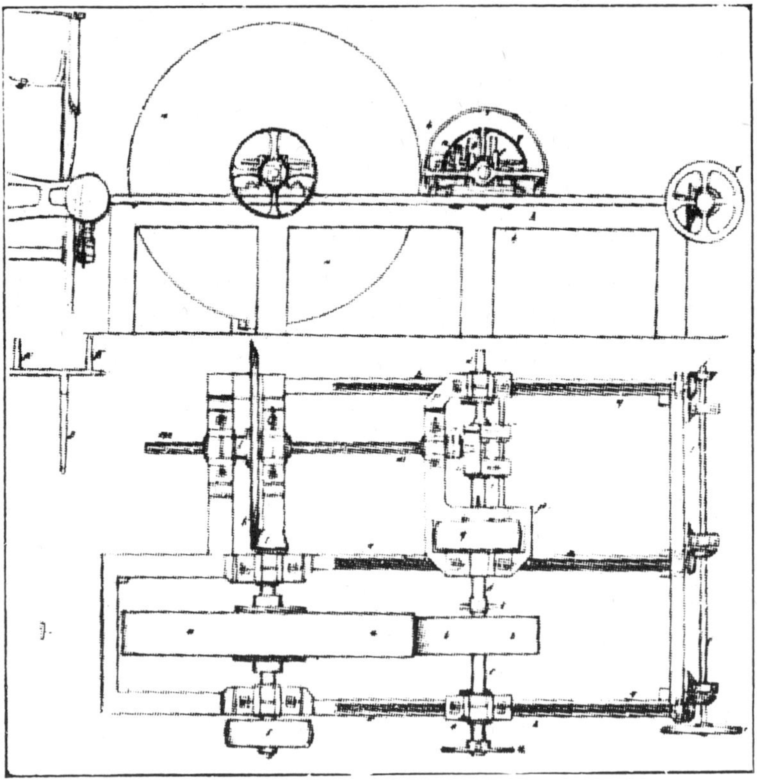

FIG. 21 WHITELAW'S GRINDING MACHINE, 1838 (*Schroeder*)

In the early history of any of the machine tools we are bound to find the name of J. C. Bodmer and his famous patent No. 8070 of 1839. In the manufacture of many of his machines he used a horizontally mounted copper disk 31 inches in diameter to do cylindrical and plane grinding of hardened iron cylinders. He later used a vertical spindle grinding machine with a copper wheel to grind the beds of lathes and other machines.

The disk-type grinding machine appears as an industrial tool in the work of James Nasmyth. The disadvantages of using emery carried on a wheel or disk of soft metal as the grinding wheel were becoming apparent, and so were the

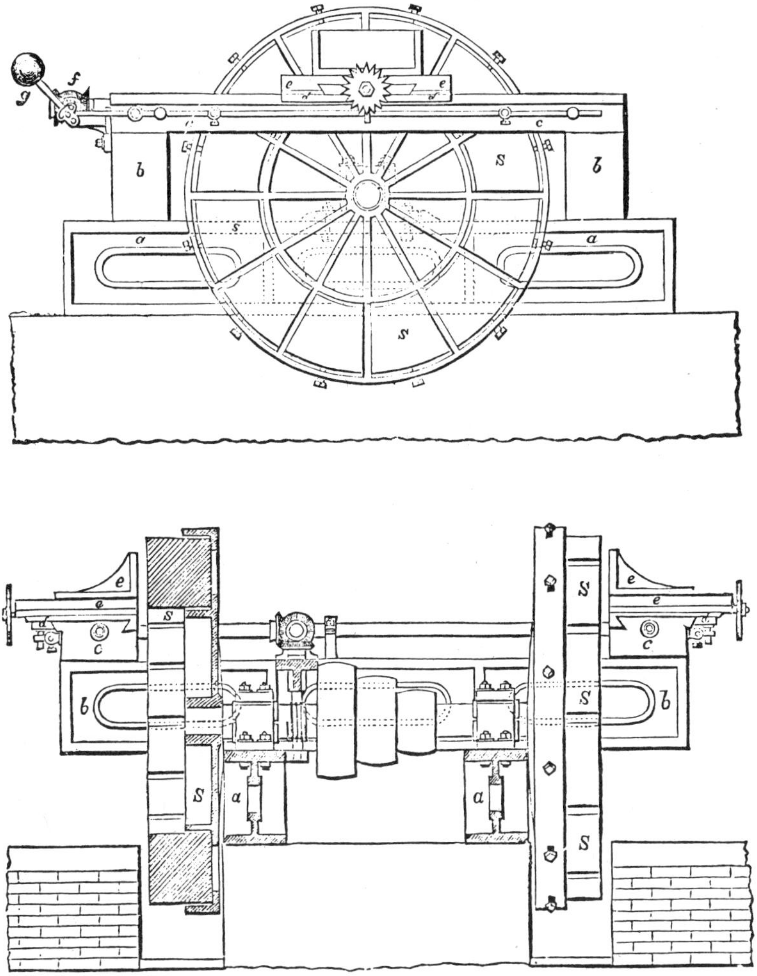

Fig. 22 Nasmyth's Disk Grinding Machine, ca. 1845 (*Byrne*)

difficulties in using large pieces of natural sandstone. Yet for industrial work, the advantages of a large wheel on a disk grinder were evident. In Nasmyth's grinding machine he used a cast-iron annular wheel about 7 feet in diameter and having twelve segments of stone about 15 inches long held in separate radial compartments by set screws in the outer circumference. (Fig. 22). This machine was dual, two such grinding disks being mounted on the same heavy foundations and driving shaft, and each having its own slide for carrying the work. Automatic feed of the work across the face of the grinding wheel was provided, as well as automatic feed into the grinding wheel.

Clearly the second generation of the great British tool builders contributed their share to the development of the industrial grinding machine.

The Vertical Grinding Machine

Although it was developed at this time primarily for grinding glass for optical purposes, the vertical grinding machine became of importance. We have already mentioned some of the earlier machines built for this purpose by Stewart and Legey, but these were rather specialized. In Hoyau's mirror-grinding machine of 1837 we see many features which were to be utilized later in the vertical grinding machine, especially those elements intended to prevent vibration and to provide precision in the control of the grinding wheel. By 1846 we find a very large and heavy rotary cast-iron table in use for this purpose, in which great care was taken with the bearings. A very solid and carefully designed vertical machine, still for mirrors, was described in 1851 by Lassell (Fig. 23), and needed only a reciprocating work table, rather than a rotary one, to give us a complete vertical spindle grinding machine.

The Surface Grinding Machine

The planer had already appeared in use by 1830, and the need for exact plane metal surfaces was evident in many machines, especially in the slides of steam engines, both stationary and locomotive. The advantages of hardening these plane surfaces were also apparent. So it is not surprising to see grinding techniques soon applied for this purpose, as well as to cylindrical surfaces. We have looked already at Stone's surface grinder of 1831.

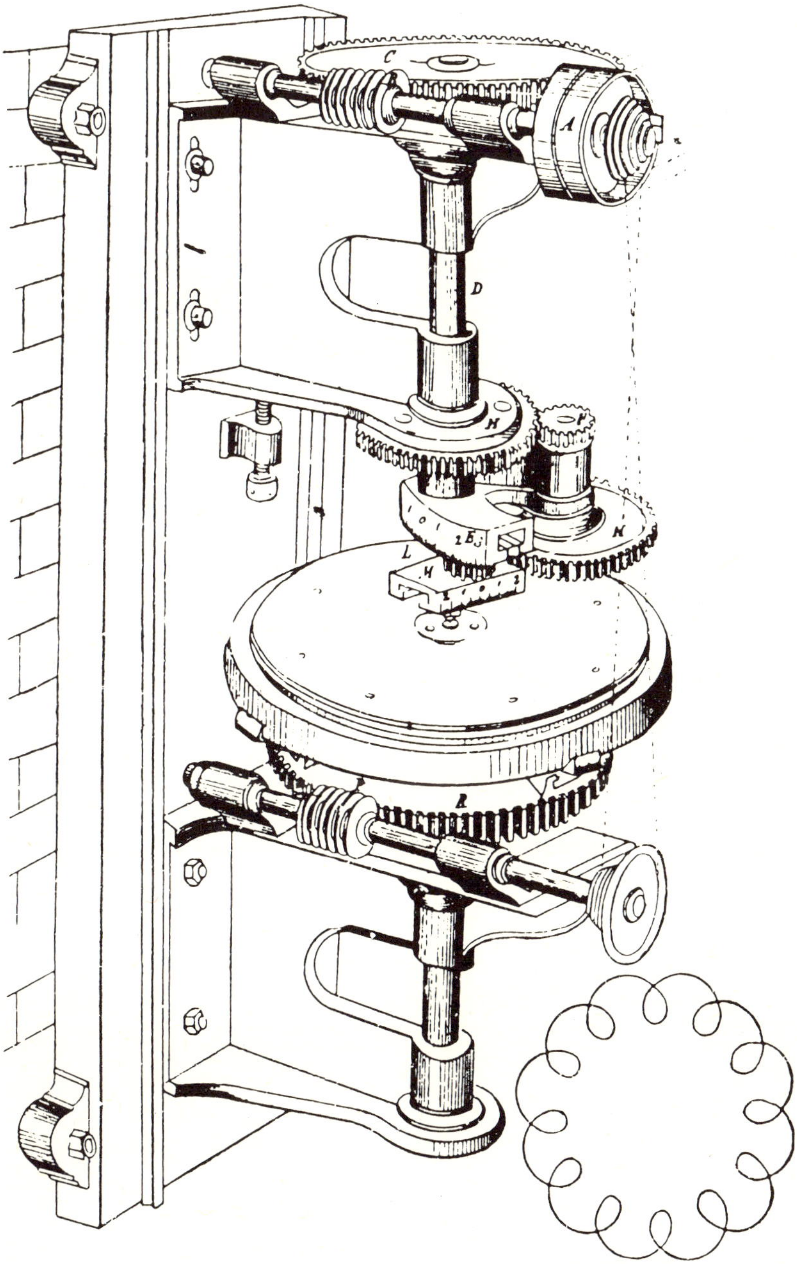

FIG. 23 LASSELL'S VERTICAL GRINDER FOR MIRRORS, 1851 (*Schroeder*)

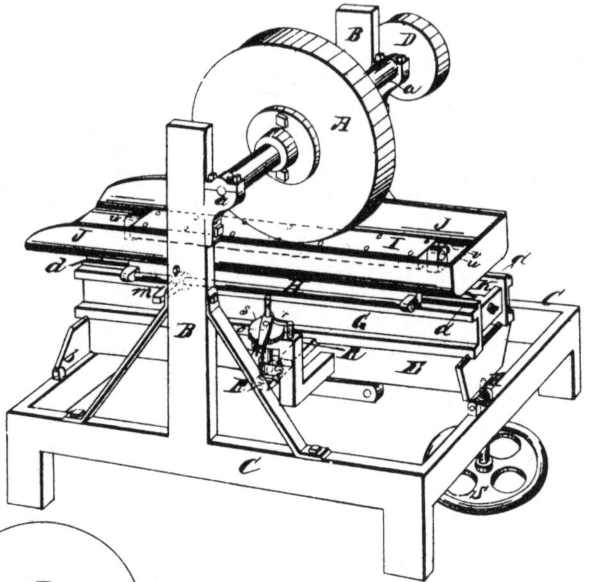

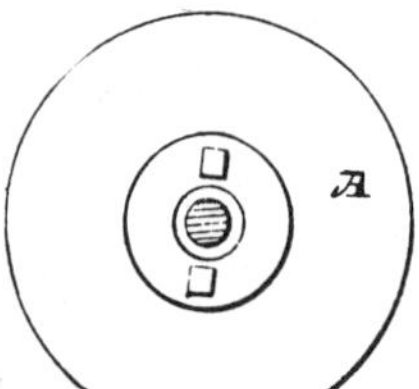

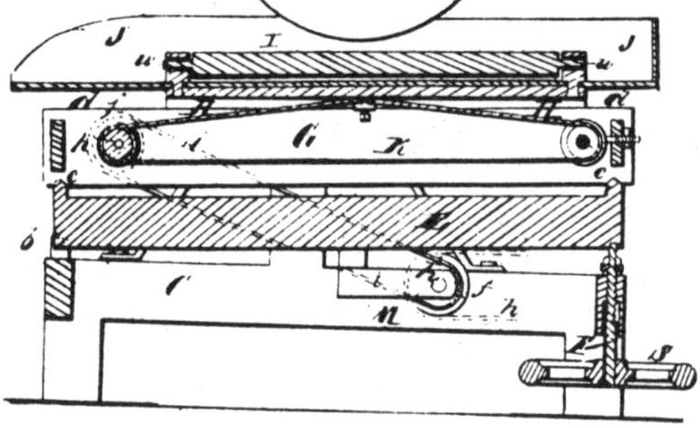

FIG. 24 DARLING'S SURFACE GRINDER, 1853 (*U. S. Patent*)

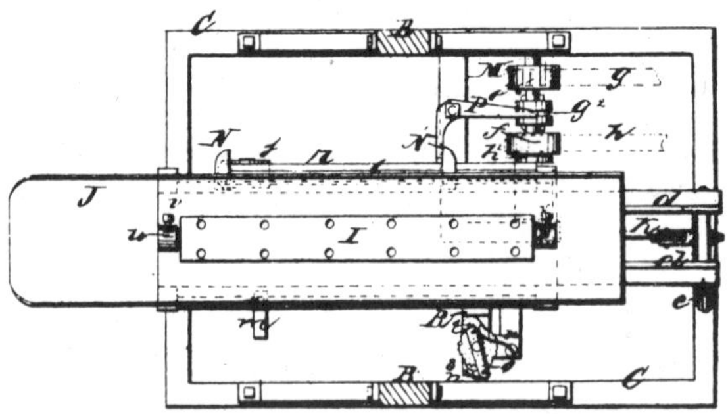

A very modern looking surface grinder (Fig. 24) was invented by Samuel Darling of Bangor, Maine, in 1853.[11] The grinding wheel was mounted on a horizontal spindle carried in bearings in a substantial frame above the horizontal work table. This table was given longitudinal feed on V-guideways by a chain drive working over two pulleys, and having an automatic reversing clutch operated by adjustable trip dogs. This assembly was carried on a saddle, also mounted on V-guides, to give the cross feed. The cross-feed assembly was supported by a rigid base hinged at one end and adjustable by a screw at the other to give vertical control of the work. The work table was fitted with a sort of trough for carrying off the water and grinding scurf. This machine was successfully used for many years both before and after Darling joined the firm of Brown & Sharpe in 1866.

The Specialized Grinding Machine

Two specialized grinding machines of this period are of interest. The first was a continuation of Leonardo's old dream of an automatic needle-grinding machine. That there had been some technical progress in this industry is shown by the drawings and descriptions of a needle factory in Dingler's *Polytechnische Journal* for 1821. But in 1858 Schleicher obtained an English patent[12] for a *self-feeding* grinding machine for making needles (Fig. 24). This is of interest because of its highly specialized character, but more so because it had automatic supply and feed of the work to the machine, and because its action was really that of a single-wheel *centerless grinder.*

Another specialized grinding machine, of great future importance, was the work in 1853, of Hiram Barker and Francis Holt of Manchester, for grinding hardened balls for

11. Darling's patent No. 9976 of Aug. 30, 1853. The machine described by Schroeder, M. Firth's British patent No. 765 of Mar. 26, 1859, is so nearly identical with Darling's as to cast serious doubt on the originality of Firth's invention.
12. British patent No. 982 of May 3, 1858.

the valves of locomotive pumps[13] (Fig. 25). These balls were ground between the grooves in two iron plates, using emery and water. It was the principle of this machine that later was to make possible the use of ball bearings to give enormous reduction of friction in all kinds of machinery.

The Grinding Machine by Mid-19th Century

Considerable advance in the grinding machine had been inspired by the classical age of machine tools after Maudslay. Power had been applied both to drive of the grinding wheel and to feed of the work. The need for strength and rigidity of construction had been recognized, and grinding machines were being made of iron and steel, with many of their own parts hardened and ground. The grinding process had become thoroughly mechanized, and was employing the basic elements of the modern grinding machine.

All the principal types of grinding machines—cylindrical, internal, and surface, in both horizontal and vertical types, and using either wheel or disk grinding—all these had appeared, together with a number of specialized machines. The various types of grinding machines were employed in industry, largely in light work, but there was also some heavy industrial grinding of rollers.

What remained to be done? First, the *mechanisms* and *precision* which had already been developed for other machine tools, especially the lathe, had to be embodied in the grinding machine. It needed particularly the high precision and rigidity of the bed on which the carriage moved. And the fine and accurate cross feed of the lathe was yet to be applied to the feed of the work to the grinding wheel.

Second, various questions of the *grinding wheel* itself—its abrasive material, its construction, and its speeds—were far from being settled. Many different types, including a few artificial wheels, had been tried out, but with only limited success. Artificial abrasives in a bonded wheel had yet to

13. British patent No. 1502 of June 20, 1853.

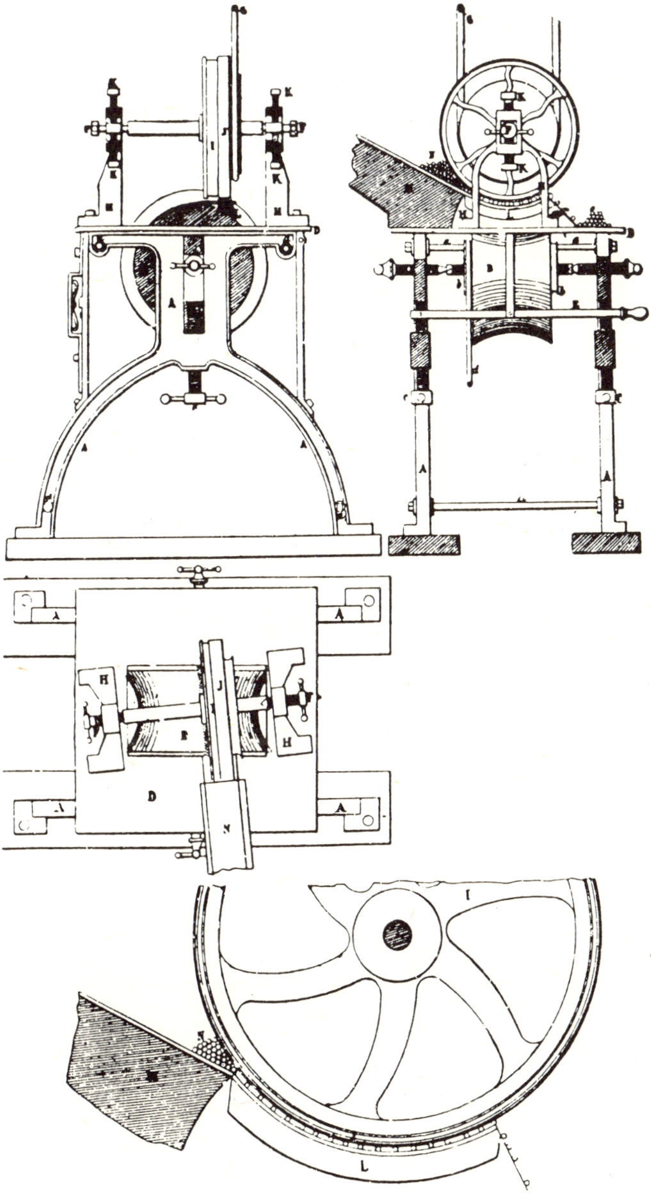

Fig. 25 Schleicher's Self-Feeding Needle Grinding Machine, 1858
(*Schroeder*)

appear, and much empirical experimental work had to be done on their use and speeds.

Third, the grinding machine had now gone through the long period of groping and trial-and-error which characterizes most machines in their early stages. It was ready to emerge in the basic, more or less standardized, forms which it was to have in the future—*when it came of age.*

Most of these trends could be foreseen in the middle of the 19th century. But the development of heavy production grinding, of automatic high-speed light grinding, and of the specialized grinding machines to meet the demands of mass production was still far in the future.

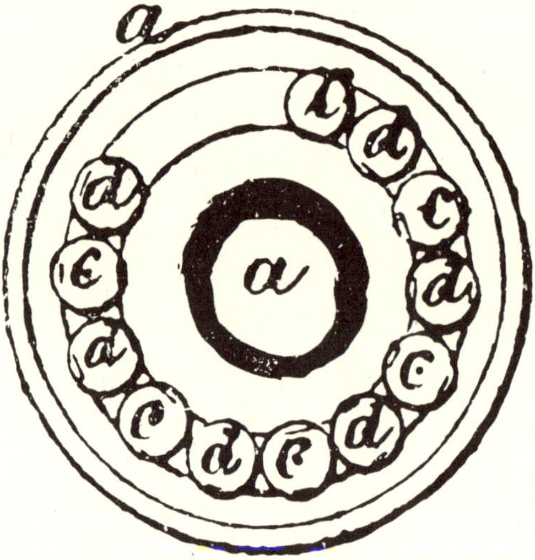

FIG. 26 BARKER AND HOLT'S BALL GRINDING MACHINE, 1853 (*British Patent*)

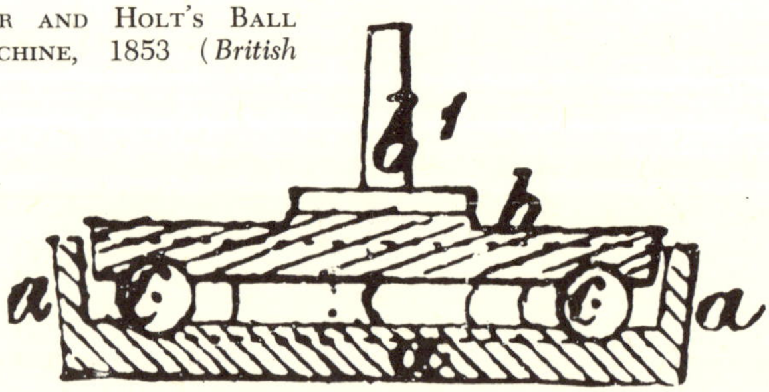

THE GRINDING MACHINE COMES OF AGE

BEGINNINGS OF PRECISION GRINDING

1860 to 1905

Webster—High Precision for Watchmaking

In the clock and watch industries the use of small grinding machines to give surface finish and precision to their very small parts had been coming into wider use since the 1820's. The exact relation of Ambrose Webster to the beginnings of the precision grinder shown in Figure 27a is difficult to establish.[1] But evidently Charles S. Moseley in early 1860 designed a grinding machine for the Nashua Watch Company, of Nashua, New Hampshire, which was brought to Waltham, Massachusetts, in 1862. It continued in use there for ten years in the plant of the American Watch Company (later the

1. In 1889 there was a patent controversy over the invention of the universal grinding machine, which involved a number of inventors, including Ambrose Webster, Mr. Moseley, and a Mr. Furbush, all of whom had invented grinding machines having some of the features of the universal grinding machine invented by Joseph R. Brown. Charles H. Norton, then of the Brown & Sharpe Manufacturing Company, investigated all these machines thoroughly and made rough drawings showing the basic features of each. (Unpublished typescript, *Investigation Relating to the Grinding Machine Patents,* with sketches, in the patent files of Brown & Sharpe, dated Mar. 25, 1889, but without initials or signature. Includes an unsigned and undated "statement of Mr. Webster of the American Watch Tool Company of Waltham, Massachusetts") Since this study was made for potential legal defence of Brown & Sharpe patents, I have followed their account and credited Moseley with the original design. On the other hand, Charles H. Norton knew Webster well enough to have him present Norton with his photograph, on the reverse of which Norton wrote, "He made a cylindrical grinding machine about 1861 for grinding the very small parts of watch machinery." Later Norton confirmed this date in a conversation in 1930. (Photograph of Webster, undated, in files of Norton Company, Worcester, Mass. Unpublished typescript, unsigned, in files of Norton Company, *Notes Taken During Mr. Norton's Conversation, April 10, 1930.* See also Norton's account of his investigation for Brown & Sharpe in his unpublished typescript *The Evolution of Grinding,* dated Dec. 12, 1929, pp. 4-5, in the files of the Norton Company.) Norton evidently believed that Webster was the original designer.

Fig. 27a Moseley's Original Watchmaking Grinding Machine of 1860
(*Norton Company*)

Waltham Watch Company). In 1872 Webster (who had founded the American Watch Tool Company, in 1860 or 1861) together with a Mr. Whitcomb, and the American Watch Company, designed a new grinder "fashioned after the one before mentioned." The new grinder replaced the old one, which was then thrown on the rubbish heap, from which it was recovered after sixteen or seventeen years and purchased by the U. S. Watch Company, where it was in use in 1889, presumably after considerable overhaul. It seems to have been actually identified as such by Webster at that time.[2]

The question of its original designer is probably best resolved by a generous bow in the direction of Moseley, and the recognition that from 1872 until 1890 he and Ambrose Webster invented a host of special grinding tools, all of which were important elements in the mass manufacture of watches by interchangeable parts. This development at the American Watch Company was not merely a change from

2. This machine is now in the possession of the Norton Company, Worcester, Mass.

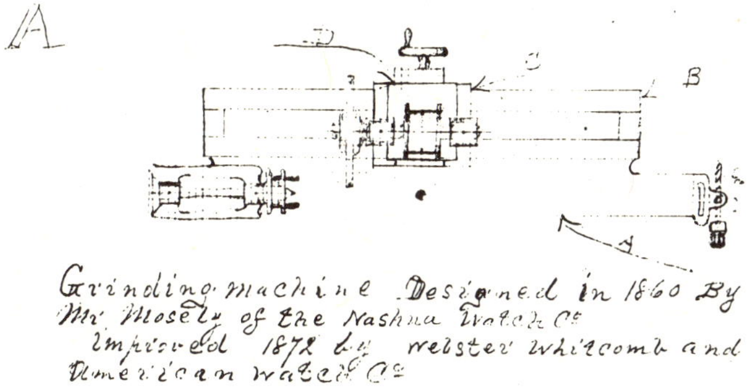

FIG. 27B BASIC ELEMENTS OF WEBSTER'S GRINDING MACHINE OF 1872 (*Brown & Sharpe*)

the old craft method of making watches; it marked the beginning of manufacture by interchangeable parts as we know it today. It was even the beginning of automation.[3]

Certainly a machine essentially like the one shown in Figure 27a was in use in watchmaking by 1860, but even by 1889 only five had been constructed. As indicated in Figures 27a and 27b, this machine did have a swivel table, but it rested on a fixed base. The grinding wheel head moved on ways to traverse the wheel along the work. The wheel itself was 3 inches in diameter and about 1/16 inch thick. It does, however, represent a substantial improvement in the cylindrical grinding machine, not only because it could do taper grinding, but because we can see that in it improvements leading to a *precision* machine tool, already made for the lathe, were here incorporated in the grinding machine. Its parts were all substantial and rigid for the work it was expected to do; they were all made of iron or steel and carefully finished

3. In the author's next monograph, History of Shop Precision of Measurement, a more detailed history and analysis of the principle of interchangeable parts as we see it in the mass production of modern industry will lead to a long needed re-evaluation of the work of Eli Whitney. For the Waltham development see E. A. Marsh, *Evolution of Automatic Machinery,* Chicago, 1896.

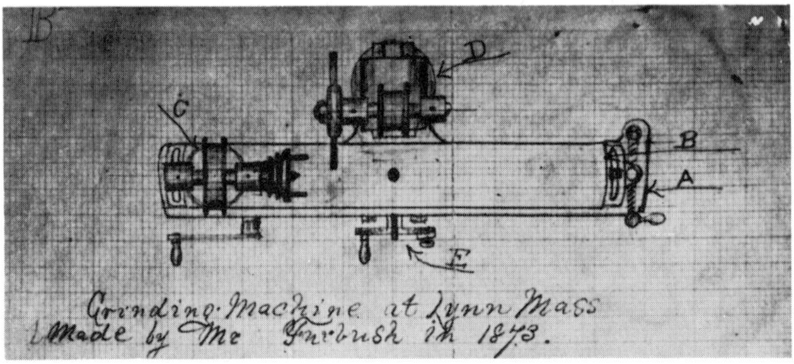

FIG. 28A BASIC ELEMENTS OF FURBUSH'S GRINDING MACHINE, 1873 (*Brown & Sharpe*)

FIG. 28B FURBUSH'S GRINDING MACHINE, 1873 (*Brown & Sharpe*)

to give precision. All controls were designed for precision work.

The Furbush machine of 1873, shown in Figures 28a and 28b, is of interest because it was built by its inventor at Lynn, Massachusetts,[4] in complete ignorance of the work of Joseph R. Brown. Yet it embodied practically all the features of Brown's universal grinding machine of 1868. It was actually built before Brown's, which was not completed or patented until several years later. Furbush's machine seems to have had no influence comparable to that of Brown's after Brown & Sharpe displayed it at the Paris Exposition of 1876. But the appearance of all these machines in the 1870's[5] does show that the time was ripe for the universal grinding machine and for the precision grinding machine to assume basically its modern form.

Poole—High Precision for Heavy Rollers

Not only was precision being sought in the 1870's for light grinding work on small parts; at this time precision was also required in some heavy machine grinding. About 1870 J. Morton Poole took a contract to make some rolls for paper mills. He expected to turn these rolls on a lathe and then to "rectify" them by putting a grinding wheel on the carriage of the lathe, a technique already well known for lighter work, such as grinding parts for sewing machines. He tried it, but the results were unsatisfactory because the finished rolls were not perfectly round. He was therefore forced to design the precision roll grinding machine.[6] His original machine is

4. Charles H. Norton in his typescript, *The Evolution of Grinding*, p. 5, says that Furbush told him that "just before the Civil War he and his partner made the patterns for this machine." Before it was built, they both enlisted and did nothing with it again until they returned. When completed, the machine proved unsatisfactory for Furbush's work in repairing shoe machinery, and it was abandoned.
5. Note also the swivel table in the patent of T. Prosser No. 65,942 of June 18, 1867.
6. Poole's basic patents are: No. 79,683 of July 7, 1868; No. 99,000 of Jan. 18, 1870; No. 104,492 of June 21, 1870; and No. 130,741 of Aug. 20, 1872.

Fig. 29a Poole's Original Precision Roll Grinding Machine, 1870
(*Norton Company*)

shown in Figure 29a and was still in use in 1931. The type is still being made and gives very accurate work.

To obtain the high precision required in these very heavy paper-mill rollers, Poole introduced an entirely new principle in machine tools. In all other machine tools, surfaces are made true by the use of guide ways and slides to determine the straight line motion of either the work or the tool, and truth of the work depends on the accuracy of the straightness of these guide ways. In Poole's machine, guide ways are used only to get the emery wheel in as straight a line as possible, but the roll itself is used to correct any errors in the guide ways.

As shown in Figure 29b, the two grinding wheels are each supported in a swing frame consisting of links which hang from shafts supported on knife edges at *H* and *J*. These knife edges are carried in small grooves on the standards S

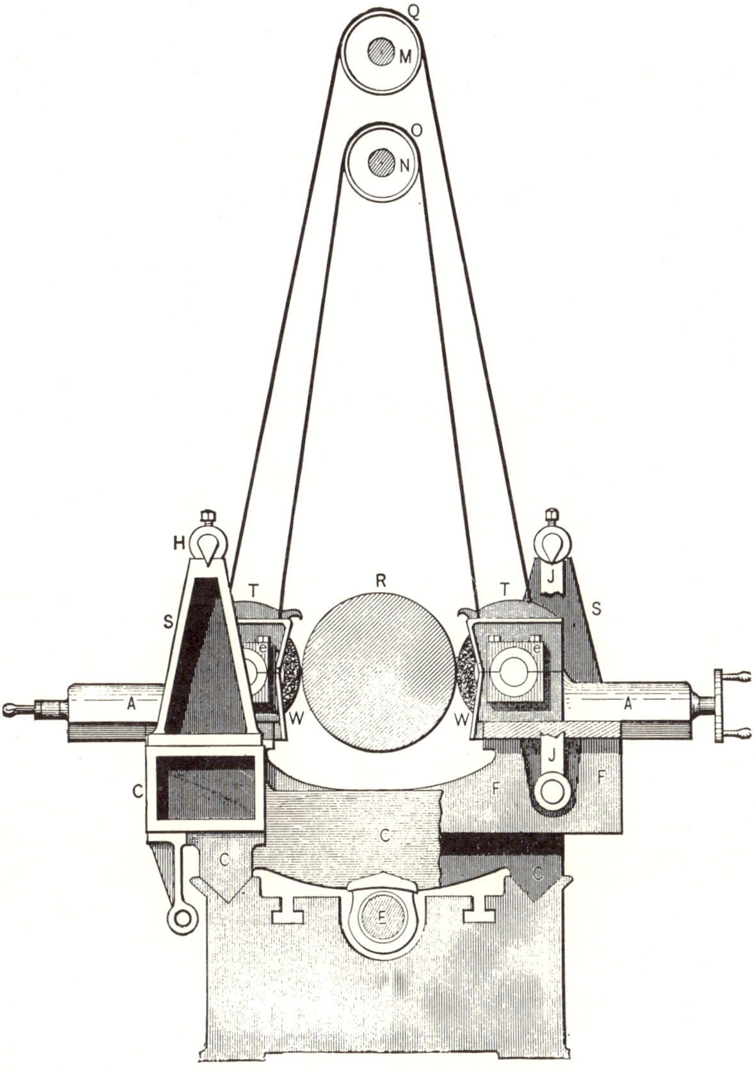

FIG. 29B BASIC PRINCIPLE OF POOLE'S GRINDER (*Rose*)

which are mounted on the carriage *C*. On top of each swing frame is a slide rest *A* which carries the grinding wheel *W*. These grinding wheels may therefore swing as pendulums in a plane at right angles to the axis of the roll. These wheels are rotated by overhead belting as shown.

The roll *R* is carried in centers as nearly parallel as possible to the guideways of the lathe and to the axes of the knife edges which support the swing frames of the grinding wheels. It is also slowly revolved. The carriage *C* is traversed along the roll on the usual guide ways by a screw *E*.

The emery wheels have been adjusted by means of the slide rests *AA* so that their circumferences are separated by the exact diameter desired for the roll when their swing frames are at the center or bottom of the pendulum swing. In this position they will not cut because of lack of feeding pressure. However, if the wheels are displaced from this center position, either by the surface of the roll being off center or by the roll being of too great a diameter, a gravitational component will force them againt the roll and they will cut until the neutral position is obtained all along the roll.

The result is a roll ground to a very high precision in straightness and uniform diameter.[7] Poole brought high precision into heavy industrial grinding and did it with more cleverness than Krupp's attempts of a generation before, for Poole's machine was perfectly feasible with the means available to Krupp. All that was required was Poole's brilliant conception.

Brown—High Precision in the General Machine Shop

There is no doubt of the importance of the universal grinding machine exhibited by Brown & Sharpe at the Paris Exposition of 1876. It marked at once the culmination of the development of the precision grinding machine up to that time, and it also introduced the basic form which the cylindrical grinding machine was to have in the future.

7. Prof. Sweet, of Cornell, tested their accuracy and found it to be within .000025 of an inch. See *American Machinist,* Aug. 7, 1886, p. 6.

However, there are a number of questions which must be carefully considered before we can get a clear picture of how and why this machine came to be and before we can give a fair share of credit to the men who brought it into being.

The introduction of the universal grinding machine was only one of many machine tools produced by the firm of Brown & Sharpe in a quest for greater precision in manufacture of metal parts, yet it grew at first out of their own manufacturing needs, as did so many of their innovations. Both the original partners in the firm were men who had a feeling for precision, as did Samuel Darling, as one time a member of the firm, and later Oscar J. Beale, their chief engineer.

Joseph R. Brown was led into precision work by both inheritance and training. His father, David Brown, was a clockmaker, and for a time after learning the trade of a machinist the son joined him in this occupation. By 1850 Joseph Brown had set up in business for himself and had built the automatic linear dividing engine which was in use at Brown & Sharpe for over 100 years.[8]

Lucien Sharpe came to work as an apprentice for Mr. Brown on September 12, 1948, and so began a lifetime association of two men whose talents so complemented each other as to produce some of the most fruitful machine tools of our day, as well as the prestige which the name of Brown & Sharpe has had ever since. Lucien Sharpe also took an interest in precision, for he had provided himself with a set of watchmaker's tools and made his own watchmaker's lathe during his apprenticeship. Throughout the long association of these two men it was Brown who was the mechanical genius and Sharpe who was the outstanding businessman, but one who thoroughly understood, appreciated, and encouraged the work Brown was doing.

The next precision machine tool constructed by Brown

8. This device and Brown's later important invention of the pocket vernier caliper and his development of the shop micrometer will be described in a later monograph on the History of Shop Precision of Measurement.

was his precision gear-cutting and dividing engine[9] of 1855. In those days, however, the principal business of the firm was still the construction and repair of clocks and watches and other small instruments of precision. But in 1858 the partners undertook the manufacture of the Willcox and Gibbs sewing machine. This device had been invented by James E. A. Gibbs in an attempt to make a crude model of a Howe sewing machine merely from a printed illustration. He finally succeeded in making his device work, only to be told that the Howe machine used two threads, and his used only one. He had invented the single-thread sewing machine. Financed by James Willcox of Philadelphia, the manufacture of these machines was contracted for with the young concern of J. R. Brown & Sharpe, which had already achieved a reputation for accuracy and high-grade workmanship. The sewing machine proved to be such a success that greater and greater expansion of the plant was required, and the manufacture of these machines soon became a matter of mass production. The profits to the firm and the demands of production then warranted the introduction of new manufacturing methods, the use of jigs and fixtures for producing interchangeable work, and the *design of machine tools primarily in order that the sewing machines could be better and more economically made*. Once invented and then offered for sale, the manufacture of these machine tools soon became more important than the making of sewing machines and the small tools.

The important characteristic of all the machine tools that soon followed from the genius of Joseph R. Brown was that they were not restricted and specialized tools only for making sewing machines, but were conceived in such broad terms that they became of first-rate importance in all machine-shop work. The first machine tool Brown & Sharpe built for sale was a turret screw machine.[10] Stimulated by the need for screw-machine products for the sewing machine, this tool became of great importance in producing many

9. See R. S. Woodbury, *History of the Gear-Cutting Machine,* Technology Press, Cambridge, Mass., 1958, p. 80.
10. This development will be discussed in a later monograph on the History of the Automatic Screw Machine.

other small parts for the needs of the Civil War. Although the universal milling machine of 1861 and Brown's formed cutter of 1864 had their first inception in other problems than the sewing machine, they too were developed in a form sufficiently general to be of wide use for many other purposes.[11]

That J. R. Brown & Sharpe had been interested in grinding problems prior to 1860 is indicated in a number of letters in their files[12] inquiring of others' experience and mentioning a grinding machine which they were building in 1858. They had even tried out an internal grinding attachment for a lathe in the same year.[13]

Joseph R. Brown's interest in cylindrical grinding grew, however, out of the needs of their sewing-machine manufacture.[14] His first grinding machines were designed for dry grinding the needle bars, foot bars, and shafts of the Willcox and Gibbs sewing machine. The basic principle of these first grinding machines was to modify a 14-inch Putnam lathe for the purpose, by adding an emery wheel on the carriage and attaching a longitudinal feed mechanism and reverse. Drawings showing the method of constructing these attachments are in the possession of Brown & Sharpe, dated respectively November 26, 1864, and November 25, 1864.

11. See the author's forthcoming History of the Milling Machine.
12. Book #9, p. 228 of Dec. 25, 1856; Book #11, p. 297 of Oct. 15, 1857 and p. 351 of Nov. 4, 1857; Book #12, p. 131 of Mar. 2, 1858 and Apr. 23, 1858. It seems possible that Darling was taken into the firm in 1866, not only to eliminate his competition in the sale of machinists' graduated scales and to get his linear dividing engine, but also to get the use of his Patent No. 9975 of Aug. 30, 1853 for a suface grinder. (Fig. 24), for it was assigned to the partnership of Darling, Brown & Sharpe.
13. Letter Book #13, p. 410 of Oct. 18, 1859.
14. For the rest of the story of the work of Joseph R. Brown on the grinding machine I have made use of the article by Luther D. Burlingame, "History of the Invention of the Universal Grinding Machine," in *Machinery,* July 1910, p. 877, and Burlingame's unpublished typescript, *The Pioneer Universal Grinding Machine,* dated May 4, 1932, in Brown & Sharpe's files. However, much original and new source material from Brown & Sharpe's files has also been used, together with material by Charles H. Norton from the Norton Company's files. The conclusions I draw as to the role of Brown in the development of the universal grinding machine are, nonetheless, my own.

A dead center pulley, an important element in obtaining precision grinding, is shown in a drawing also of 1864, but it is clear that cylindrical grinding had been done before this, for there is also a drawing of September 22, 1862, for a solid back rest. Later a spring-back rest consisting of a series of leather washers carried on a spring rod was used. Skill in adjusting these back rests was required to prevent chattering. In these early days wet grinding was done by holding a wet sponge against the work. Later a supply of water was provided through a nozzle, as shown in a drawing of December 22, 1866. By 1867 a guard was placed on the ways of the lathe to keep off the water and the emery.

About thirty of these grinding lathes were built in the next few years; some were retained for use in making the Willcox and Gibbs sewing machines; others were sold for $400 to $450 each to various sewing machine and other manufacturers both here and abroad.[15] They were listed in the first Brown & Sharpe Catalogue of May 15, 1867.

These grinding lathes were quickly copied at home and overseas since they were found to be adapted to a great variety of work, such as "sizing reamers (straight and taper), arbors, bushings, rings, plugs for templates, and many other articles." They were commonly sold with the Slate taper attachment and with the ways and gearing protected by guards. Because of their lack of rigidity it was usual to add a heavy weight to the carriage, as shown in Figure 30. Nevertheless, great skill and patience were required to produce accurate work. So much skill was required (at least part of it consisted of knowing where to put one's foot on the floor, or when and where to rest a hand on the machine to get the desired results) that the operator of the first machine at Brown & Sharpe, Thomas Goodrum, was in 1867 receiving the then princely wage of $7.00 a day. The dignity associated with such skill was made apparent to the public by the fact that he wore a tall silk hat to work. In fact, in a photo-

15. See Brown & Sharpe sales records July 28, 1865, to Dec. 4, 1882. This firm has in their possession an incomplete example of these grinding lathes. For other manufacturer's machines see *American Machinist,* Sept. 4, 1890, p. 1 and Sept. 22, 1892, p. 3.

Fig. 30 Grinding Lathe, About 1867 (*Pratt & Whitney*)

graph taken of the employees and the old wooden shop on South Main Street, Providence, in 1872, just before it was abandoned, there are three men shown wearing tall silk hats—Joseph R. Brown, Lucien Sharpe, and Thomas Goodrum!

The grinding lathes proved unsatisfactory from a production point of view. So much time and skill were required to do satisfactory work on them that the work kept piling up ahead of the workers. There were also many other kinds of work best done by grinding, but which could not be done on the grinding lathe. For several years Brown had under consideration building a machine designed from the floor up as a grinding machine. He had a number of ideas of what features such a machine should embody, but he clearly did not think them out in detail himself. In 1868 he "dictated a drawing for a Universal Grinding Machine to Mr. G. W. Spaulding. This drawing embodied Mr. Brown's ideas, at that time, as to what was wanted in such a machine."[16] Dated April 10, 1868, it was laid aside because the factory was too busy to undertake the work at that time. This drawing, not Brown's own, shows a plan and elevation of a swivel table, mounted on the reciprocating table, and provided with a tangent screw for adjusting the angle between them. There is a swivel for the headstock, but none for the wheel slide by which one could grind steeper tapers than with the swivel table alone. Furbush's machine of 1873 had embodied all these features, but of course we do not know how many of them were included in the machine for which he had made patterns before he went off to the Civil War. At any rate, it seems very unlikely that Brown and Furbush had any influence on each other. Furbush's machine came to nothing; Brown's went on to become the pattern of the modern grinding machine.

The projected universal grinding machine was picked up again briefly in November and December of 1873, but it was not until April of 1874 that work was commenced in earnest and drawings were completed. Actual construction began in June of 1874. The exact influence of Joseph R. Brown in this period is not clear. None of the drawings made at this time are his own; some were done by Mr. Parks and by Mr. Lewis. The actual construction was done by Mr. Phillips.[17]

While he was employed at Brown & Sharpe, Charles H.

16. *Investigation Relating to the Grinding Machine Patents*. The drawing is still in Brown & Sharpe files.
17. These drawings are all still in Brown & Sharpe files.

Norton was told by Mr. Parks that "Mr. Brown left merely sketches, that he, himself, made the design." At any rate, Norton wrote, "You may not know that Mr. Parks really designed the Universal grinding machine in use when I first went to Brown and Sharpe." [Nov. 1886][18] Since Norton had later to redesign the machine to eliminate defects which he believed would not have been in it had Brown, and not Parks, made the original design, we may be sure that at least Norton believed that, while Brown had had the basic concepts of the universal grinding machine, the actual design was unfortunately not his.

The first machine was nonetheless completed in time to be shown in the Brown & Sharpe Catalogue of January 1, 1875. It included now a swiveling wheel stand and a cross-feed hand wheel which was also on Furbush's machine and apparently on the grinding lathes although no record of such use can be found. Also included with the first machine was an attachment for internal grinding, shown on the floor at the right in Figure 31. A drawing of this device dated 1875 is in Brown & Sharpe files. The final design also included means of carrying off the water used in grinding. One interesting feature shown in the drawings of this machine is that the spindle for the wheel had its bearings and boxes of hardened steel having double tapers which would require a universal grinding machine to produce them. The tapers are such that they could not have been made on a grinding lathe with a taper attachment, nor could they have been done on Webster's machine. How were they made? On a machine sired by Wheaton's of 1834?

By 1880 at least fifteen of these machines had been sold in the United States and abroad to manufacturers of various types of machinery, and with most satisfactory results.[19]

18. Charles H. Norton, *The Evolution of Grinding*, in Norton Company files. Also letter of Charles H. Norton to William A. Viall, Vice President, of June 9, 1936 in Brown & Sharpe files.
19. See Brown & Sharpe sales records, Jan. 3, 1876, to Jan. 20, 1880. Also testimonials printed in Brown & Sharpe Catalogue for 1889, p. 22. Original letters are in Brown & Sharpe files.

FIG. 31 ORIGINAL BROWN & SHARPE UNIVERSAL GRINDING MACHINE OF 1876. (Now on display at their plant. The patent date on the door is for the Bement patent on the internal bracing shelves in the frame.) (*Brown & Sharpe*)

The universal grinding machine was not patented until after the death of Joseph R. Brown.[20] Its influence on later grinding machines may be seen by a comparison of Figures 31 and 32. The modern machine has many improvements, especially in rigidity, in the controls, and in the use of electric motors, but it is still basically the universal grinding machine as Joseph R. Brown conceived it.

The influence of the universal grinding machine in industry is best expressed in the words of one well qualified to know from personal experience. In a manuscript in Brown & Sharpe files Henry M. Leland, of Cadillac and Lincoln automobile fame, who was a foreman at the Brown & Sharpe works during the years of development of the grinding machine, wrote:

20. Patent No. 187,770 of Feb. 25, 1877.

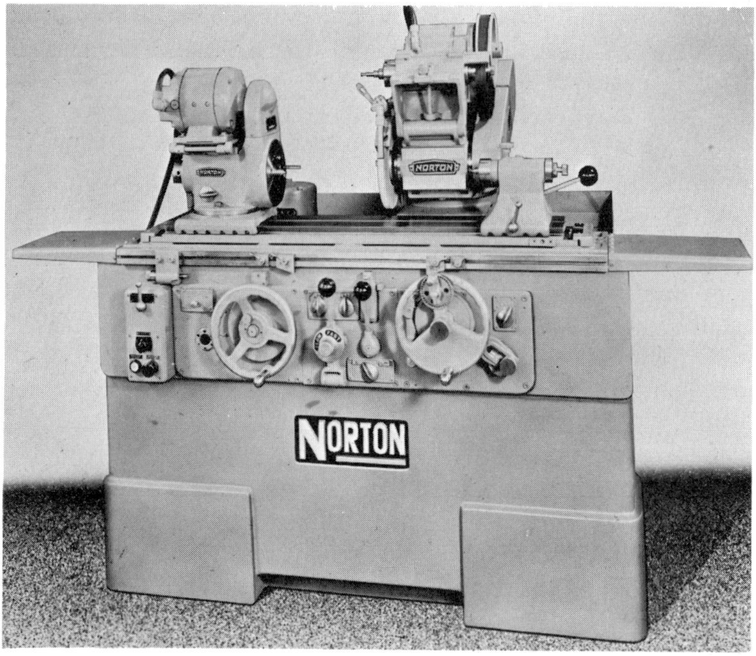

FIG. 32 A MODERN UNIVERSAL GRINDING MACHINE, 1957 (*Norton Company*)

"What I consider Mr. Brown's greatest achievement was the Universal Grinding Machine. In developing and designing this Machine he stepped out on entirely new ground and developed a machine which has enabled us to harden our work first and then grind it with the utmost accuracy at the same time protecting the ways—the surface on which the platen travels—from emery and grit; also the improvement of revolving the work from dead centers thus eliminating the error of live spindles and dead centers. If all these machines should be suddenly taken away, it is hard to imagine what the results would be. It would be impossible to make any more hardened work for the best parts of our machinery and tools, that would be round, true, and accurate in every detail to the closest possible limits. This in my judgment is one of the most remarkable inventions and too much cannot be said in its praise, or in acknowledgment of Mr. Brown's perseverance, wonderful initiative and genius . . . I know of none who deserves a higher place or who has done so much for the modern high standards of American manufacture of interchangeable parts as Joseph R. Brown."

Once the notion of obtaining precision grinding by applying to the grinding machine the precision mechanical elements already developed for other machine tools had been understood, it took only a short time for the other types of grinding machines, whose early history we have already examined, to appear in precision form.

In 1867 the Seth Thomas Clock Company was using a grinding wheel mounted on a planer as a surface grinding machine to work dies and other hardened stock. By March 1, 1877, the Brown & Sharpe catalogue offered a well-developed planer-type surface grinder (Fig. 33a). In their catalogue for February 1, 1883, is listed a post-type surface grinder (Fig. 33b), basically like those in common use for light work today. In early 1884 this machine was being built with automatic traverse of the table. By October 1, 1885, they were also offering a surface grinder (Fig. 33c) in which the work table was mounted on a movable knee and the axis of the grinding wheel remained fixed. However, the wheel was capable of cross feed. A remarkably well-developed face-type surface grinding machine (Fig. 33d) was shown in their catalogue of January 1, 1887. This same catalogue also offered a planer-type surface grinder with an 8′ x 2′ x 2′ capacity. Heavy surface grinding, principally for locomotive and stationary steam engine parts, had arrived.

Brown & Sharpe sales records show that at least twenty-five of these various surface grinders were sold from February 7, 1883, to November 22, 1886, again to customers both at home and abroad.

The increased use in industry of precision cylindrical grinding led Brown & Sharpe to bring out a plain cylindrical grinding machine, listed in their catalogue of 1877. The demand for a somewhat larger and heavier universal cylindrical grinding machine was met in their "Large Size" machine using an emery wheel 18 inches in diameter and 1 inch thick (Fig. 34). It is listed first in their catalogue for February 1, 1883. In 1887 Charles H. Norton redesigned the original universal grinding machine to give it greater rigidity.

FIG. 33A PLANER-TYPE SURFACE GRINDING MACHINE, 1877 (*Brown & Sharpe*)

FIG. 33B POST-TYPE SURFACE GRINDING MACHINE, 1883 (*Brown & Sharpe*)

FIG. 33C KNEE TYPE SURFACE GRINDING MACHINE, 1887 (*Brown & Sharpe*)

FIG. 33D FACE TYPE SURFACE GRINDING MACHINE, 1885 (*Brown & Sharpe*)

The first adequate internal grinding attachment put on the market was the one listed in the Brown & Sharpe catalogue of July 1, 1880. Between 1887 and 1889 Charles H. Norton perfected a spindle which made a really satisfactory internal grinding machine possible. Many of their cylindrical grinders were sold with only the internal grinding attachment.

FIG. 34 LARGE UNIVERSAL GRINDER, 1883 (*Brown & Sharpe*)

By 1885 grinding processes had become so widespread, and the difficulties of training workmen in the quite different technique of grinding so troublesome, that Brown & Sharpe put out their first *Treatise on Grinding Machines.* This book went through many editions; its contents are a measure of the general state of grinding at the time. The first edition, of October, 1885, described the various grinding machines and emery wheels in some detail, and then took up such topics as: plain grinding; wet grinding; centers; finish grinding; internal, angular, and side grinding; sharpening of reamers and taper and side-tooth cutters; and, most significantly, the use of gages. For it was about this time that *precision machine tools* came together with *shop precision of measurement* to produce results of enormous importance to industry. The book also listed the then recognized advantages of grinding: *a*) could be used on both hard and soft work, *b*) was cheaper and saved time, *c*) gave a high order of accuracy and quality of finish. Brown & Sharpe gages were offered as an example of what could be done.

By 1890 the standard types and the principal elements of the precision grinding machine had been well established; grinding methods were widely used in a number of light industries and were spreading into others. Further development of the light grinding machine centered on the control of the feed mechanism of the machine. The grinding lathe and the original universal grinding machine had been fitted with automatic power longitudinal feed, as had most of the surface and face grinders. Automatic power cross feed was introduced by Charles H. Norton[21] in Brown & Sharpe machines of 1897. Hydraulic feeds were not long in appearing. For hydraulic longitudinal feed Brown & Sharpe used in 1903 the invention of A. G. Cassidy[22] on its larger machines. Another hydraulic control for both longitudinal and cross feed was developed by Parks and Thatcher[23] of Brown & Sharpe in 1905. This had the feature in the cross feed of a quick motion to move the wheel rapidly away from the work for about half an inch to allow a new piece of work to be inserted. It then moved the wheel back quickly to the grinding position, thereby saving considerable time on each operation.

About the turn of the century, controls of this sort, a convenience and a saving in the light grinding machines, were becoming a necessity on the *heavy production grinding machines* being introduced by Charles H. Norton, who had left Brown & Sharpe and gone to the Norton Company to bring about the revolution in industrial production grinding.

21. His basic patents are No. 590,850 of Sept. 28, 1897, and No. 591,357 of Oct. 5, 1897.
22. His patent No. 734,221 of July 21, 1903.
23. Their patent No. 797,216 of Aug. 15, 1905.

THE GRINDING WHEEL

TRANSITION FROM NATURAL ABRASIVES TO THE ARTIFICIAL WHEEL

1820 to 1910

The Problem of the Grinding Wheel

In order to bring about modern production grinding a number of elements were needed. In this chapter we shall be concerned with those connected with the grinding wheel itself. Thus far in our study we have given the grinding wheel only passing attention; it now becomes crucial for production grinding, either on the heavy machines to be developed by Charles H. Norton, or on the complex high-production centerless grinders of nearly a generation later. The early wheels of naturally occurring stones or the use of emery would not do. What was required was a wheel which could withstand high centrifugal forces and hold the form of its grinding surface under production conditions. For both the heavy cylindrical and the centerless grinding machines, large and wide grinding wheels were required. But it was more important to make available *grinding wheels of known and uniform characteristics,* in order that a body of at least empirical data might be accumulated on speeds and types of wheels to be used for different metals and kinds of work. The naturally occurring stones could not meet these conditions, nor could the use of emery. Wheels capable of only very light grinding and lasting only for a brief period were usually constructed by the machinists themselves even in Joseph R. Brown's day. To be sure, various abrasive grains had been bonded into a number of types of matrices to produce an artificial wheel, but even men of long experience could seldom get any uniformity by this technique. Emery also had its limitations as an abrasive. The solution required for *production* grinding wheels was to use artificial abrasives bonded into a number of matrices of fairly well-known and controllable characteristics.[1]

Emery, Corundum, and Metallic Wheels

To the end of the 18th century all grinding was done with one of three naturally occurring abrasives. Emery, an impure aluminum oxide containing 20 to 25 per cent of iron oxide and about 5 per cent silica, was used in granular form and embedded in the surface of wheels of soft metals or of wood, often leather covered. Because emery occurs with an irregular structure and mixed with widely varying impurities, it has seldom been used as a solid wheel of the stone as quarried. Until 1856 when deposits were found in Chester, Massachusetts, it was obtained principally from deposits in Asia Minor and especially from the Greek island of Naxos.

Sandstone is an abrasive of quartz crystals bonded by nature with silica, iron oxide, or similar materials. With the exception of a few machines for grinding spindles for textile machinery and some use in grinding gun barrels, it has not been used in grinding machines, because it also varies widely in its abrasive character and its structure. However, it has been the usual abrasive in the familiar hand grinding wheel for sharpening and similar work.

Diamond dust, although clearly in use for grinding and cutting gems by 1694, had only special and very limited use in grinding until the 20th century.

The use of emery grains carried by wheels of wood, leather, or soft metal was adequate for polishing or light grinding operations in which little metal had to be removed, only a good surface finish was desired, and precision was of little importance. It had also limitations on its cutting ability.

In 1825 the first specimens of natural corundum were brought from India to England. This material, an almost pure aluminum oxide, proved to have better cutting proper-

1. In this chapter and the one that follows I have used some information obtained first-hand from the people concerned, as well as from the records of the Norton Company, by Mildred McC. Tymeson for her company history, *The Norton Story,* Worcester, Mass., 1953.

ties than emery. Although at first corundum had to be imported from India, Ceylon, Burma, and Madagascar, large deposits were found in 1871 in North Carolina and Georgia. Still later, corundum was discovered in Canada.

Well down to 1860 emery and corundum were used in granular form embedded in a softer surface, as in a lapping wheel. With the demand for greater precision and the appearance of true grinding machines, a grinding wheel of greater cutting capacity and greater precision was needed. The first solution was to make a wheel of cast iron, carefully turned and balanced on a lathe. These wheels varied from 3 to 6 inches in diameter and from ⅛ to ¼ inch in thickness. Then glue, of the common sort used by woodworkers, was dripped on the outer circumference as the wheel slowly turned in the lathe, in order to distribute the glue evenly over the surface. Sometimes dovetails were cut in the iron surface to hold the glue in place against centrifugal force. Abrasive grains, usually emery, were then sifted over the surface of the glue, and the whole allowed to dry. The resulting emery wheel was then trued with an oilstone.[2] These wheels were usually made by the machinists themselves. Of course they did not last very long, but they were used on only very light cuts. There was much discussion of the merits of using several sizes of abrasive grains, but in any case, the cutting qualities of these wheels of necessity varied considerably. These wheels had their limitations, but because of the troubles with the early solid emery wheels, they held their own in practice until about 1885,[3] although solid wheels had been in use in 1864 on the Brown & Sharpe grinding lathes.

The Solid Wheel—Emery or Corundum and Bonds

When corundum was first brought from India to England in 1825, word came that it was mixed with gum resin in Mala-

2. Charles H. Norton, letter to William A. Viall, dated June 6, 1936, in Brown & Sharpe files.
3. *Idem.*

bar to form a grinding wheel used on a very simple machine for hand grinding of gems.[4] These wheels were made of two-thirds corundum cemented by one-third gum resin. The mixture was heated to a paste, kneaded, and rolled on a stick to give uniform consistency. The paste was then flattened and worked into wheel form by a rolling pin, and finally allowed to harden. A hole was then melted in it for mounting the wheel on a shaft.

In the following generation a rash of inventors and manufacturers tried out all combinations of all possible substances in the effort to perfect a satisfactory solid grinding wheel in Germany, England, France, and the United States. By 1837 solid bond emery grinding wheels were being made in England.[5] They were being sold commercially in France by 1843 and in Germany by 1850.

Out of the many kinds of bonds tried for solid emery or corundum wheels a few principal types had emerged by 1900: vulcanized rubber, glue and cements, silicate, vitrified, shellac, and frit. Resinoid bonds did not appear until the 1920's and were seldom used on grinding machines.

Vulcanized rubber as a bond for abrasives was described by Deplangue[6] in 1857. This type of grinding wheel was invented independently in the United States in the same year by T. J. Mayall.[7] His wheels were manufactured by the New York Belting & Packing Company prior to 1865, for they were in use at the Seth Thomas Clock Company when Charles H. Norton first went there in that year, and were sold under the name of Vulcanite. Improved forms of rubber wheels are in use for various grinding purposes today,

4. Dingler's *Polytechnische Journal,* 1825, p. 523. Also reported in the United States in *Journal of the Franklin Institute,* 1826, p. 346, and *Farmers, Mechanics, Manufacturers, and Sportsmans Magazine,* N. Y., July 1827, p. 190.
5. Samuel Parbly in *Civil Engineers and Architects Journal,* Nov. 1838, p. 378.
6. See his British patent Mar. 28, 1857. These wheels were being manufactured by Warne & Company in London in 1862. Schroeder says that the American vulcanized rubber wheels were in use in Europe in this same year.
7. See his U. S. patents No. 25,747 of Oct. 11, 1859; No. 25,841 of Oct. 18, 1859; and No. 125,600 of Apr. 9, 1872.

especially for cut-off wheels and as regulating wheels in centerless grinders.

From 1867 until 1872 a number of types of solid emery grinding wheels were made of various substances mixed with the abrasive and held together by glue or cement of one sort or another. Patents were issued for wheels using concrete lime, zinc oxide, granite and kaolin with mica, barium carbonate with zinc oxide and chloride, paper pulp, and boracic acid. None of them amounted to much, with the exception of Vulcanite, which was still being made as late as 1915. The manganese oxychloride bond of 1876 was to be revived much later and is used today, especially in dry disk grinding.

The first silicate bond for a grinding wheel is that of F. Ransome[8] in 1859. He mixed granulated emery or powdered glass with silicate of potash or soda to form a plastic mass. This was given the desired shape and then dried and heated so that the abrasive was held in a silicate bond. Sometimes he added some clay, and the wheel was heated to "a bright red heat," so this wheel may have been vitrified. But Ransome also said that he did not heat to the point where the ground glass fused. In 1867 the English engineer Donkin tested the metal-cutting capacity of this type of grinding wheel and found it to be more than fifty times that of the best Newcastle natural stone. The silicate bonded wheel began in the United States with the work of Gilbert Hart[9] in 1872. Hart had been employed in making artificial sandstone for window sills. In this process sand was mixed with silicate of soda and a clay, dried, baked, and chemically treated to resemble sandstone. Hart applied this technique with success to bonding solid emery wheels.

The vitrified bonded wheel began earlier, with the patent of Henry Barclay in 1842.[10] He took various parts of

8. See his British patent No. 2929 of June 21, 1859.
9. See his patents: No. 201,778 of Mar. 26, 1878, No. 226,066 of May 30, 1880, and No. 228,257 of June 1, 1880.
10. See his patent No. 9337 of Apr. 30, 1842. Also the account in *Journal of the Society of Arts*, London, Mar. 22, 1878, p. 362, which claims an English origin for the solid emery wheel but credits an American development.

Stourbridge clay and emery, and pressed the mixture into molds. The wheel was then fired to "a low white heat." But "cracks and distortions in firing" led finally to abandonment of the process. Commercial vitrified wheels were being made in 1872 using a borax composition, but the company failed because of financial difficulties.

In the following year Swen Pulson, a potter in the shop of Franklin B. Norton, of Worcester, Massachusetts, was present at a discussion of the merits of Hart's artificial grinding wheels. He announced that he could make a better one—of clay. He first tried a mixture of emery, clay, and the slip clay used for glazing bean pots. When his first wheel was fired, it melted and ran. In his second wheel he used too much clay, and it blistered when fired. But on the third attempt Pulson got it right. The pottery shop showed little interest in Pulson's wheels, so he set up in business for himself with some success. In the next few years, however, he wandered from one company to another perfecting his clay process and having a part in the early development of the Vitrified Emery Wheel Company, the Sterling Grinding Wheel Company, the Norwich Emery Wheel Works, the Springfield Emery Wheel Company, and finally the Abrasives Materials Company, of Philadelphia, where he served as superintendent for twenty years. He had in the early years returned at intervals to work at F. B. Norton's shop. During one of these intervals his process was patented by Norton, as was the custom for employers in those days (No. 187,167 of February 6, 1877), but using feldspar in place of the slip clay. By 1880 Norton had one of his six employees working full time making emery wheels—John Jeppson. As the business of manufacturing solid emery wheels grew by leaps and bounds, Frank Norton's health declined, and he sold the business to the Norton Emery Wheel Company, incorporated for the purpose in June of 1885. So began the Norton Company of today.

In the ten years after Norton's patent of 1877 other inventors tried various other bonds in attempting to get a vitrified wheel. Two other bonds have been used, but seldom in wheels for grinding machines. In 1880 Henry Richardson, of the American Watch Tool Company, Waltham, Massachusetts, originated the shellac, or "elastic," solid grinding wheel. Although Richardson was interested in a grinding wheel for the fine work of watch-tool making, it came to have wider uses, especially for cut-off wheels. The frit type of wheel comes a bit later, about 1905, and was brought out by the Norton Company primarily for grinding cut glass.

Grades and Speeds

By 1885 the elements necessary for a solid grinding wheel for production work were known, but much trial and experimentation were required before really satisfactory wheels were available. The actual technical advances made from 1885 to 1910 in the manufacture of vitrified grinding wheels were kept as trade secrets and are difficult to reconstruct today. Probably most of them depended upon the individual judgment and skill of foremen and key workmen.

Soon after the Norton Emery Wheel Company moved into its new plant in January, 1887, the volume of business was so great that "puddling" of the mixture of clay, feldspar, and emery had to be done by a machine driven by a small engine. The resulting mix was scooped out by hand into a steel ring on a plaster bat. After drying, the puddled wheel was then "shaved" to approximately correct dimensions, until 1906 on a potter's kick wheel.

The goal aimed at in these early vitrified wheels is indicated in Norton advertising of the time. Their "porous and open texture" was said to result from the fact that the wheels had not been "tamped or subjected to any pressure whatever." The wheels were also declared to contain nothing but "cutting properties, and that they will run equally well wet or dry. Free from dust or smell." Lack of smell was an incentive to machinists used to glue in their emery wheels.

The kilns for firing the wheels were vertical and round, with a down draft, and heated by egg coal. Temperatures were measured by little, partly glazed clay ring "pyrometers" which changed color in proportion to the temperature. If these rings were broken, the taste of the clay would also indicate how the firing was going.

It was soon found that during firing some wheels became slightly saucer-shaped, and that such wheels invariably broke in use. They were therefore trued to make their two sides parallel. The wheels were then ready for grading.

John Jeppson's personal skill was crucial here. After tapping the wheel sharply to produce the ringing note that indicated a sound wheel, he dug a metal tool very much like a screwdriver into the wheel to determine its grade of hardness. At first these were no standard grades of hardness; each wheel was chosen for a certain kind of work. Jeppson said later, "It became a natural thing to line up soft to hard on an alphabetical scale. Our first grading standards were the stubs [used-up wheels] brought back from satisfied customers."

In 1884 an extensive series of tests was undertaken by Gilbert Hart, of the Detroit Emery Wheel Company, to determine the relative efficiency of emery and corundum. Corundum was conclusively shown to be far superior. By 1893 emery had largely given way to corundum, soon to succumb to the artificial abrasives for production grinding wheels.

Such were the beginnings of standardized grinding wheels, but the cumulative empirical knowledge gained did make possible tables of grain size and wheel hardness suitable for all kinds of grinding work.[11]

When Charles H. Norton first went to work for Brown & Sharpe in 1886, he was soon assigned to solve the problems they were having with grinding in their own shop. With the assistance of Charles L. Allen, of the new Norton Emery Wheel Company, he discovered two causes for the troubles: *a*) the solid abrasive wheels then in use were not in perfect balance,[12] and *b*) the surface speed was too great for the grades of wheels in use.

11. See Brown & Sharpe, *Treatise on Grinding Machines,* Providence, R. I., 1891, pp. 87-89, 142-147.

12. After he went to work for the Norton Company he invented a machine for giving accurate dynamical balance to grinding wheels.

With reasonably uniform grinding wheels available by this time, the problem for all practical grinding men faced by a specific grinding job was: What size of grain do I want? What grade of wheel—hard or soft? And what surface speed shall I use? The practical machinist wanted an easy simple solution, preferably a table which would give him a quick answer.[13] The experiments of Norton and a few others indicated, however, that the problem was more complex. In fact, even though the results of modern abrasive research give a sound answer, it is based upon a complex analysis of a given grinding situation, and is even now only a "scientifically educated guess," frequently subject to modification in the light of practice.

Norton published a series of articles in the *American Machinist* on the relation of wheel grades and speeds to the work to be done.[14] In these articles he refused to be drawn into the glib rules desired by "practical machinists" who thought Norton knew more than he was telling. However, given a *specific* grinding problem, Norton not only stated the proper choice of wheel and speed, but gave in full the analysis of the various factors by which he had arrived at his solution. But Norton was doing more than accumulating empirical knowledge of the use of grinding wheels; he was also taking the first steps in the attempt to understand the nature of abrasive cutting of metals. By microscopic examination of grinding chips he was able to show that most grinding wheels then in use put most of their energy into heat to melt the metal. By a proper choice of wheel and speed the grinding process became one which *cut* the metal into curled chips, very small but otherwise like those of the other metal-cutting tools.[15]

By 1905 there was available reasonably satisfactory information on how to use the artificial grinding wheel in prac-

13. See Brown & Sharpe, *Treatise on Grinding Machines,* Providence, R. I., 1891, pp. 89-90.
14. *American Machinist,* 1865, pp. 431, 451; 1901, p. 717; 1903, pp. 185, 305, 353, 641, 794, 1575; and 1904, p. 58.
15. Brown & Sharpe and others were working along similar lines. See *American Machinist,* 1903, p. 942.

tice. It was no accident that Colvin and Stanley's *American Machinist Grinding Book* appeared simultaneously in New York and London in 1908, and James J. Guest's *Grinding Machinery* was published in London in 1915.

Dressing, Truing, Mounting, and Safety

As grinding became more of a precision operation and as artificial wheels came into use and went to speeds setting up high centrifugal forces, a few smaller problems of grinding had to be met.

In the maintenance of a grinding wheel it is desirable that a fresh supply of sharp cutting points of the abrasive grains come into contact with the work as the ones in use become dull. In short, a grinding wheel, like any other cutting tool, has to be "resharpened" or dressed at intervals. In an ideal wheel the strength of bonding would be such that, when the wheel is properly selected and applied, abrasive grains are continually released from the bond under normal grinding pressure just as they begin to become dull, and fresh, sharp ones take their place. This ideal grinding condition is seldom achieved in practice, and even today grinding wheels are too often improperly used. As a result "glazing" or dulling of the cutting grains, or loading or smearing of the wheel face with the material being ground occurs. The machinist decides that the wheel, not himself, is dull and proceeds to dress or sharpen it. The "practical machinist" invariably uses this operation too often and too heavily, and thus wastes good abrasive wheels, when more careful choice of wheels and speeds for the work in hand would reduce this dressing to a minimum. Dressing does, of course, have a place in preparing the wheel face for rough grinding cuts.

These dressing devices have nearly always consisted of a series of hardened steel points arranged to rotate about an axis parallel to that of the grinding wheel. When forced against the rotating wheel, they crush the dulled abrasive grains free of the bond to expose new cutting points. The first of these devices, for dressing the sandstone grinding wheel, was patented by C. Altzschner, of Dessau, in 1860.

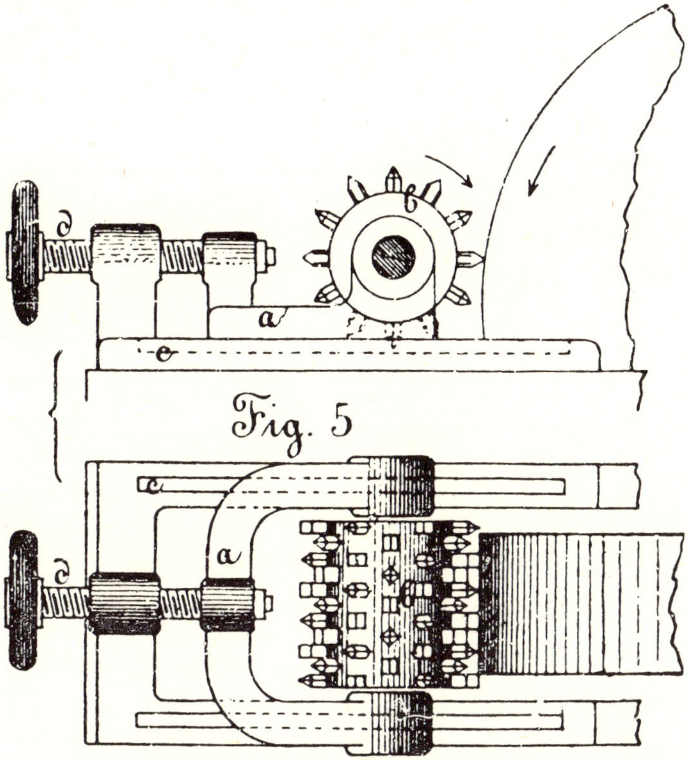

FIG. 35 PARTICK'S WHEEL DRESSER, 1860 (*Schroeder*)

Shortly thereafter appeared the improved dressing tool of J. E. Partick, of Chemnitz, shown in Figure 35. The familiar Huntington dressers were used as early as 1865. Other more elaborate types were developed by 1877.[16]

It is also necessary to keep the cutting surface of a grinding wheel running true and of the shape desired—this process is called truing.[17] A device for this purpose was con-

16. Unpublished, unsigned typescript, *Notes Taken During Mr. Norton's Conversation,* dated Apr. 12, 1930, in the files of the Norton Company. See also Brown & Sharpe Catalogue, Mar. 1, 1877, p. 21, and Appleton's *Cyclopedia of Applied Mechanics,* New York, 1882, Vol. II, p. 73.
17. Not to be confused with the truing of the *sides* of the wheel at the factory in order to be able to balance it properly, even though the technique is very similar.

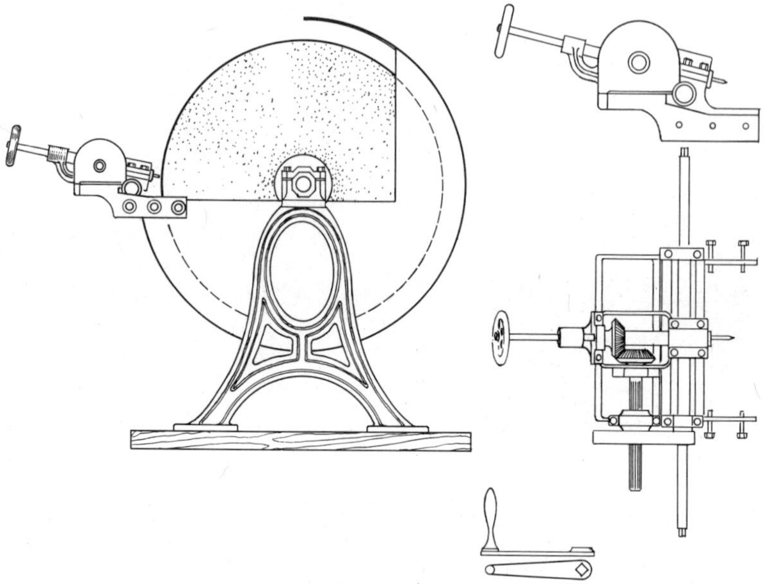

FIG. 36 SAMMANN'S TRUING DEVICE, 1859 (*Schroeder*)

structed by A. Sammann in 1859. As will be noted in Figure 36, Sammann had the axis of the truing point at a small angle to the radius of the wheel.

The need for careful truing of the grinding surface of the wheel had been well recognized in 1890. This operation was generally done by means of a diamond-point tool held in the hand or in a fixture.[18] The later method is shown in Figure 37. By 1903 a convenient device for truing a wheel for simple form grinding of a semicircular groove was available.[19]

The basic principles of dressing and truing grinding wheels were set forth by Charles H. Norton[20] in 1905. He pointed out that the practice of "truing" a wheel with a piece of another grinding wheel at best only dressed the wheel,

18. Brown & Sharpe, *Treatise on Grinding Machines,* Providence, R. I. 1891. p. 95.
19. *American Machinist,* 1903, p. 489.
20. *American Machinist,* 1905, p. 142.

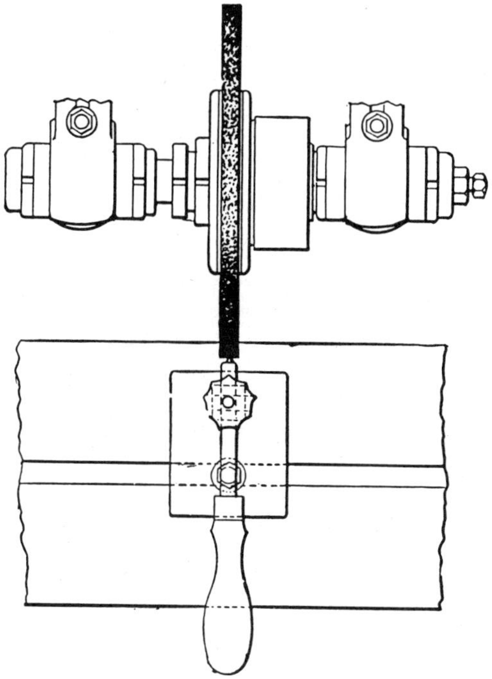

Fig. 37 Diamond-Point Truing of a Grinding Wheel, 1891
(*Brown & Sharpe*)

and that dressing was in fact not really necessary for precision work properly done, but applied only to the crude grinding done in foundries. However, for precision work truing is of the first importance. For this work there is no substitute for the diamond point, but it must be mounted in a tool post, not held in the hand, and fed slowly past the revolving wheel and well flooded with water. This truing cut need normally be of only a few thousandths of an inch.

Diamond-point truing of the wheel became by 1925 of even greater importance. We then have machines in which a diamond gaging point is set opposite to the wheel. With this point linked to a dial, and once set up with a master gage, the operator knows he can grind until the dial reads zero. Then he must true his wheel by using a diamond point with a micrometer adjustment to save wear on the wheel.

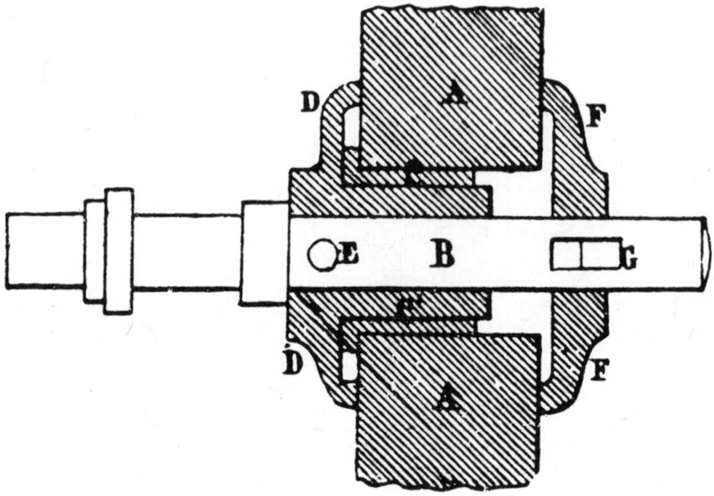

FIG. 38 PHILIPP'S WHEEL MOUNTING, 1863 (*Schroeder*)

Some machines are set up to true the wheel automatically at regular intervals, to ensure a correct wheel and good cutting at all times. These machines usually compensate in their automatic cross feed for the resulting change in the size of the wheel.

There was some serious interest by 1863 in the problem of mounting the grinding wheel on its spindle. The method shown in Figure 38 is that of Philipp. This method long antedates studies on the stresses in grinding wheels, later instigated by the bursting of high-speed artificial wheels.

Brown & Sharpe advocated by 1891 "elastic washers placed between the wheel and the flanges. Sheet rubber is best for this purpose, but soft leather will answer very well. In some cases, manufacturers of emery wheels attach a thick, soft paper washer to each side of the wheel for this purpose, in which case no further attention is required in this direction."[21] This method of wheel mounting is shown in Figure 39.

21. Brown & Sharpe *Treatise on Grinding Machines,* Providence, R. I., 1891, p. 95.

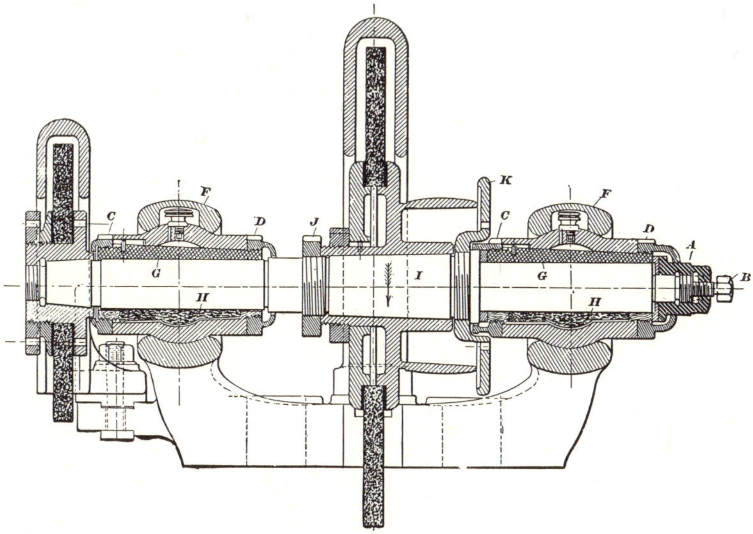

FIG. 39 WHEEL MOUNTING AND WHEEL GUARDS, 1891 (*Brown & Sharpe*)

Abrasive Industry reported in 1925 that, after carelessness of the workman, the principal cause of wheel bursting was improper mounting—improper, unrelieved flanges; excessive tightening of flanges; and in dry grinding undue expansion of the wheel by heat, especially in cup wheels.[22]

The early solid grinding wheel gave a great deal of trouble by bursting, and grinding was recognized as a hazardous occupation. The early accidents were in many cases caused by the weakness or imperfection of the wheel itself, as well as by unfamiliarity or carelessness on the part of the workmen. Attacks on the problem of wheel safety were made from several directions. One was to hold the wheel in special flanges which would take the strain set up by the centrifugal force. One company sold concave safety collars to fit the convex sides of a "safety" wheel almost out to the working edge. The outward forces were expected to tend to wedge the wheel between the flanges. Sets of flanges in several diameters were provided for use as the wheel wore

22. *Abrasive Industry,* Sept. 1925, p. 279 and Oct. 1925, p. 307.

FIG. 40 TESTING GRINDING WHEELS, 1900 (*Norton Company*)

down.[23] Another firm had a similar arrangement, but used stepped wheels and corresponding steps on the flanges.[24]

Another approach to the problem was careful testing of the wheels at the factory. The Norton Company[25] took the lead and tested all their grinding wheels at speeds at least twice their normal working speeds. The machine they used is shown in Figure 40. Two wheels could be tested at a time inside a heavy iron guard. The counterbalanced covers permitted two wheels to be loaded while two others were being tested. The results of the tests were sworn to daily before a

23. *American Machinist,* Feb. 1, 1894, p. 3.
24. *American Machinist,* 1896, p. 1028.
25. *American Machinist,* 1896, p. 638 and 1900, p. 84.

justice of the peace and were indicated on the label of the wheel before shipment.

A series of experimental tests[26] of the bursting speeds of solid grinding wheels was made by Charles H. Benjamin of the Case School of Applied Science in 1902. These showed considerable variation in the bursting speeds of the commercial wheels tested—from 2¼ to 3¾ of their normal working speeds. Benjamin's tests showed that, while a higher factor of safety might be desirable, the causes of bursting of wheels were to be sought in the machine shop rather than in the factory which made them. Benjamin believed that because the velocity was greatest at the rim of the wheel, so was the stress; but he did not fail to note that the cracks all radiated from the lead bushings at the center.

It seemed evident that wheels were going to burst anyhow. They still do. The answer then seemed to be better training of the workmen in their care and use, plus wheel guards. These had been fitted to grinding machines before. Brown & Sharpe machines[27] had them in 1891 (Fig. 39). A number of other types of wheel guards were invented,[28] but the simple iron hood has survived.

Artificial Abrasives— Silicon Carbide and Aluminum Oxide

By 1895 emery had been replaced for most grinding by the better cutting properties of corundum. But although corundum of the quality required for grinding wheels had been discovered in a number of places in the world, it was scarce and expensive. Like emery, it varied in its characteristics. What was needed to perfect the grinding wheel was an abrasive which could be made cheaply out of plentiful and

26. *American Machinist,* 1903, p. 1421. A similar set of tests was made in Germany about the same time.

27. Brown & Sharpe, *Treatise on Grinding Machines,* Providence, R. I., 1891, p. 16.

28. For a French guard using sliding chains see *American Machinist,* 1905, p. 605. Various types of 1908 are shown in Colvin and Stanley, *American Machinist Grinding Book,* New York, 1908, Ch. VIII.

easily available materials, and an *abrasive of uniform and controllable characteristics,* preferably as hard as the expensive diamond.

The first of the new artificial abrasives was silicon carbide, better known under the trade name given by the man who developed it, carborundum. It had been produced as a laboratory curiosity in mid-19th century, but it was discovered independently in 1891 by Edward G. Acheson, who first put it to technical use as the first artificial abrasive which does not occur in nature at all.[29] Acheson was experimenting on the reduction of iron ores with natural gas when he noted that clay objects in his furnace were impregnated with carbon and became much harder. He had for a number of years been searching for a method of crystalizing carbon to produce synthetically the diamonds used as abrasives. It seemed likely that with the higher temperatures of the electric furnace this process might be carried further. He described his first experiment:

> "The first experiment was with a furnace constructed of an iron bowl lined with carbon, in the central cavity of which was placed a mixture of carbon and clay; through the mixture was passed an electric current of sufficient amount to fuse the whole mass—a violent reaction following the fusion—the iron bowl, and a rod of carbon suspended in the centre of the mixture, forming the electrodes. After the mass had cooled, it was removed, broken and carefully examined, when a few bright crystals, blue in color and apparently very hard, were found to be in that part which immediately surrounded the carbon electrodes. They were exceedingly small and only served to convince me that more and better arranged experiments would produce the desired results."[30]

At first Acheson did not know the composition of these remarkable crystals and so named the new substance "carborundum," from "carbon" and "corundum." Soon identified

29. An excellent exhibit of Acheson's work and his process may be seen in the Smithsonian Institution, Washington, D. C.

30. E. G. Acheson, "Carborundum: Its History, Manufacture, and Uses," in *Journal of the Franklin Institute,* Sept. 1893. The account later given in his autobiography, *A Pathfinder,* New York, 1910, differs in a few unimportant respects.

as silicon carbide, it proved to have the greatest hardness of any abrasive then in use, except the diamond.[31]

	Knoop Scale
Hard Steel	740
Garnet	1350
Tungsten Carbide	1880
Corundum and Aluminum Oxide	2000
Silicon Carbide	2500
Boron Carbide	2800
Diamond	7000 plus

Acheson was one of the first engineers to recognize the importance of production grinding:

> "Perhaps no other use to which it [carborundum] can be put will equal in importance that as an abrasive material, and should it find no other, this alone would be sufficient to class it as one of the most valuable of the materials used by the artisan. The use of emery and corundum in the form of wheels and special shapes, while of comparatively recent introduction, has grown to wonderful proportions. At no period in the world's history has the value of time been more highly appreciated than at the present day. Economy of time, increased output with a given amount of labor, and the resultant cheapening of production and consequent lower selling price of the article produced, are the demands of the times . . . . It is probable that the introduction of no other single tool into our factories, mills, and shops, has produced so great a saving in labor as the emery wheel . . . . The amount it [carborundum] will do and save will depend wholly upon its hardness and fitness as an abrasive, over and above these qualities as manifested in emery."[32]

A company was soon formed to develop the uses of this product. The ingredients were silica sand, coke, a little sawdust and salt, and tremendous amounts of electric power. At first the silicon carbide was so expensive that it was sold as a substitute for diamond dust and at comparable prices. But by November of 1893 the *Scientific American* could report: "The cost is found to be not more than half as much as that of mining and preparing corundum." By 1895 the Carborundum Company was prepared to produce this new abrasive on a large scale; they therefore moved their plant to Niagara Falls, New York, to have available the cheap electric

31. See Acheson's patent No. 492,767 of Feb. 28, 1893.
32. Acheson, "Carborundum", *Journal of the Franklin Institute,* Sept. 1893.

FIG. 41 SILICON CARBIDE FURNACE IN OPERATION (*Norton Company*)

power of the hydroelectric plant just completed there. Three men, Acheson, Frank J. Tone, and F. A. J. Fitzgerald, developed the furnaces[33] and the techniques of commercial production. It proved possible in these furnaces to produce silicon carbide crystals of uniform quality and of various desired characteristics by careful control of the raw materials and of the operation of the furnace (Fig. 41). At first the crystals of silicon carbide produced had a leaf shape and therefore had relatively few cutting points. It proved possible, however, to produce crystals more grain shaped and with more cutting points.

Carborundum grinding wheels appeared commercially as early as 1896.[34] Because of the hardness of silicon carbide these wheels proved to be especially suitable for grinding the hardest materials. But silicon carbide has a brittle nature and is therefore not suitable for some types of metal grinding,

33. See Acheson's patent No. 560,291 of May 19, 1896.
34. The catalogue of the Sterling Emery Wheel Manufacturing Company for 1896 offers them on order.

because the cutting crystals break off before they become dull. This condition was partly remedied by making these wheels more dense. This required "pressing" the wheel rather than "puddling," a technique developed by Arthur Malm of the Norton Company in 1911. The production grinder today could not do without silicon carbide wheels.

Artificial corundum is more than merely corundum made artificially on a commercial scale. Artificial corundum, like silicon carbide, could be made of relatively cheap materials, with the use of large amounts of electric power, and the product was also of uniform and controllable characteristics. In fact, control of the chemical composition and the manner of crystalization permit desired variations of the properties of artificial corundum within rather wide limits, which makes it for many grinding purposes the preferred abrasive.

Artificial corundum also has a trade name, alundum, given it by the firm which has been most important in its development, the Norton Company. It, too, had been produced on a laboratory scale prior to its independent discovery and development by Charles B. Jacobs[35] in 1897.

Jacobs' first work was done in the electric furnace laboratories of the Ampere Electro-Chemical Company, of Ampere, New Jersey. His process consisted basically of crushing, calcining, and then fusing a high-quality bauxite (nearly pure aluminum oxide) with small quantities of coke and iron borings. The fusion was carried out in intermittent electric arc furnaces, at temperatures of about 3700°F., in which the bauxite was fused into crystals and the impurities reduced to their metals.

After Jacobs' invention of the process, lack of power and other facilities led to a transfer of the operation to Rumford Falls, Maine, where larger scale experiments were made in 1899 and 1900. The Norton Emery Wheel Company secured the rights to the process and in 1901 opened the first production plant at Niagara Falls, New York. By 1906 aluminum oxide was to replace emery and corundum completely

35. Jacobs' U. S. patent No. 659,926 of Oct. 16, 1900.

Fig. 42 Aluminum Oxide Furnaces in Operation (*Norton Company*)

in Norton grinding wheels; in fact, this was one of the reasons for changing the name of the company to simply the Norton Company.

One important basic problem remained in the manufacture of aluminum oxide—the furnace. The early production furnaces were simply large pots, reënforced with heavy steel. They were lined with blocks of carbon whose joints were filled with a carbon paste. This lining gave a great deal of trouble and sometimes allowed the molten mass to flow through to the furnace shell. The resulting hot spots had to be watched for and a hose played on them promptly. Aldus Higgins of the Norton Company noted that the water treatment seemed to congeal the fused mixture just enough so that it acted as its own insulator. This idea led him in 1904 to the invention of the water-cooled furnace (Fig. 42), which cut the cost of making aluminum oxide in half.

In 1908 the Norton Company began experiments using bauxite of extra high purity. By increasing the furnace temperature they soon produced their white 38 Alundum. Wheels of this abrasive were on the market in 1910, replacing for some purposes their brown regular Alundum abrasive.

It then becomes clear that by 1910 the practical grinding machine operator had available to him the advantages of artificial abrasives in artificial solid grinding wheels, of a number of types and grades of uniform and known characteristics. He had also adequate information on the use of these wheels in his grinding machine to do the precision work at hand. But Charles H. Norton had twenty years earlier seen in the future development of grinding wheels and grinding machines the far greater possibilities of precision production grinding.

CHARLES H. NORTON

AND

HEAVY PRODUCTION GRINDING

1890 to 1920

Genesis of the Revolution—Brown & Sharpe

Charles H. Norton had his first experience in light grinding while in the employ of Seth Thomas Clock Company.[1] He went to work for Brown & Sharpe in November, 1886, where he very soon was involved in problems of production grinding. Their universal grinding machine had been found to embody all the mechanical motions that Joseph R. Brown

1. In addition to other sources, I have made use in this chapter of Charles H. Norton's unpublished typescript, *The Evolution of Grinding,* dated Dec. 12, 1929, as well as a long letter which he wrote to Mr. Howard W. Dunbar of the Norton Company, dated Dec. 6, 1929, in which he speaks very frankly of his work and its influence. I have also used the brief unsigned, unpublished typescript, *Notes Taken During Mr. Norton's Conversation,* dated Apr. 12, 1930. All these documents are in the Norton Company's files. All quotations and references in this chapter are from these documents unless otherwise indicated. Since at the time they were written, Mr. Norton was 78 years of age, he is not always quite correct in his dates. However, the correct dates are easily established by many references to him and to his work in trade journals, such as *Machinery* and the *American Machinist.* In these documents Norton's principles are somewhat clearer in his mind than they appear in his articles published at the time, but even at Worcester his employers were a little afraid to lose the grinding wheel business of Brown & Sharpe, so that these later accounts, being private, at least give Norton's actual sentiments. Norton's published articles divide into three series. The first were summaries of grinding practice of the day as advocated by Brown & Sharpe (*American Machinist,* 1897 pp. 337, 374, 394; 1898, p. 181). The second were very tentative comments on grades and speeds of wheels for various kinds of work, based upon empirical principles discovered with some trouble. The third series includes his own theories of production grinding (*American Machinist* 1903, p. 1486), and finally culminates in his paper Dec. 8, 1910, before the A.S.M.E. Seminar on Grinding. Papers were also given on this occasion by representatives of Brown & Sharpe, and of Pratt & Whitney. H. M. LeLand and others were also present and commented.

had provided for in the drawings of 1868, but the actual machine, as designed largely by Parks, proved to be less than adequate, even in their own shops. Norton was assigned to find out where the trouble lay and how to correct it. Although he was given a year, if necessary, to do the job, within a few months Norton found that the cause of the complaints and the expense of the work was the fact "that the machine was so light and its construction so frail, that it would be impossible to do work, *commercial work,* with any success, except by great skill and great patience . . . . they attempted to use the machine as a *production machine* . . . ." [*Italics mine*] In addition to some troubles with the wheels in use on it, Norton found a number of technical defects in Parks' design, especially the design of the spindles, which could not be run tight enough to do good work. But Norton had taken the first step that was to lead him to the concept of precision production grinding; he had recognized the need for much heavier construction in grinding machines.

Brown & Sharpe authorized Norton to redesign the universal grinder, which he did, giving it the proportions which it still retains. But even this early he ran into the prejudices that were to plague his struggle to make production grinding a reality in industry. Even on the redesigned machine, which had a wheel only one-half inch thick, the workmen would not use the full width but continued to true off all but a little sharp corner, "so that in practice Brown & Sharpe did not increase production . . . . Their production remained the same but their quality improved." Norton now had the second element of production grinding: larger and wider grinding wheels.

Although Norton had some experience in the design of specialized, production grinding machines in 1888-1889, when he designed a machine to grind the triple cylinders of the Westinghouse air brake,[2] it was the years 1890-1896 which he spent in Detroit as a partner in the firm of LeLand, Faulconer, & Norton Company[3] which gave him first-hand

2. See his patent for the telescoping spindle No. 429,697 of June 10, 1890.
3. Later the Cadillac Automobile Company. Henry M. LeLand was the leading figure in the development of the Cadillac and at the forefront

experience in production machine tools and an early interest in the technical problems involved in making automobiles.

When Norton returned to Brown & Sharpe in 1896, business was slack. In order to keep himself occupied he undertook the design of a heavier plain grinding machine. After much argument he finally got the firm to make six in the first lot. He had been told that it must not be any heavier than the Universal. When it came out some 150 pounds heavier, Norton was called on the carpet and lectured. Mr. Richmond Viall, then the Superintendent, objected that there was no need for such a machine. Even when Mr. Bach of the F. E. Reed Company, makers of lathes in Worcester, said that such a machine would be ideal for grinding the spindles of lathes, Mr. Viall still objected. Norton persevered and designed a larger machine[4] to meet this demand and also to grind the overarms for milling machines, but it "was a very frail affair. I did not dare to make what I had in mind. I was cautioned I must not have a big wheel, I must not have a wide wheel, because wide wheels were absolutely impossible; you could not grind work smooth with them. So as I remember it, I ventured to put a wheel 1 inch wide on . . . ." Norton's ideas were beginning to take form.

The idea of using much larger and wider wheels, led naturally to the concept of plunge grinding, in which the wheel is not traversed along the work, but is fed in to it, the width of the wheel being of the dimension and form which are desired in the part to be ground.

> Right at this point in my experience there I conceived of the ideas that have been worked out here in the Norton Company since. In 1897, the latter part of the year, I asked for a conference; there I made a suggestion that wheels as wide as 1 foot on the face would eventually be used, and that we would not traverse the work except when it was so long that the expense of

---

in the development of the manufacture of automobiles in Detroit. Faulconer was, in 1899, the first to design a machine for production grinding of hardened bevel gears for bicycles.

4. Shown in *American Machinist,* 1900, p. 636. The author has been unable to establish any clear connection between the experience gained with the Brown & Sharpe "Large size" Grinder of 1883, shown in Figure 34, and Viall's opinions on heavy grinding machines and on large and wide wheels.

> machine and expense of wheel would make it impractical. Or when the work was so small in diameter, in proportion to the length, as to make it impractical. But in all other cases we would use very very wide wheels and not traverse the work. That suggestion nearly caused a riot. I cannot tell my feelings when I realized that I was being ridiculed; I do not know how to express myself. That conference broke up very soon."

But Norton was also advancing to a fourth element in precision production grinding:

> "A month or two later I asked for another conference in which I said that if a machine was made heavy enough, and the mechanism was heavy enough, we could make an actual accurate micrometer out of the machine itself, and when the operator moved the index .00025 it actually reduced the piece .00025". That is something that had never been done, up to that time any way, and no one believed then it could be done. I was told that such a suggestion of a micrometer machine was ridiculous."

The idea of using the precision of the machine itself as a means of obtaining the precision of which grinding is inherently capable was now clear in Norton's mind.

The heavy machine, the much larger wheels, and the grinding techniques which Norton proposed to use of course required a much more *powerful* machine as well. The Bliss Press Company of Brooklyn, New York, wanted a very tough and hard toggle pin for the clutch of their presses. Norton carried out a number of experiments which showed that "if I had a machine powerful enough, heavy enough, it could be done." But this fifth proposal was also turned down as impractical.

The sixth element of Norton's revolution was the idea of saving time and labor cost (the most expensive item in nearly all manufacture) by not taking the finishing cut on a lathe. After the rough turning of a part, the practice was then to finish cut on the lathe to within .002 of an inch of the required dimension. Since this finish cut was taken with a very fine feed, it took a great deal of time. Meanwhile the lathe operator sat idly by merely watching it. The part was then ground to final dimension. Norton wanted to rough turn on a lathe, and then finish the part to the required dimension and surface directly by grinding, using heavier cuts than were at that time common. The reaction to this important idea was that grinding is a refinement. Refinement must of necessity cost more. You cannot get something for nothing. Norton was unable to get across his point that he did not expect to get something for nothing. He expected to pay for refinement with the higher cost of the heavier machine and the greater power it needed. This greater power and expense were to be expended over a shorter time. Labor was the highest expense and likely to become more so. Therefore a substantial overall saving would result. "This proposition was turned down as impractical. It would wear out the wheels too fast,[5] it would chatter and burn—a lot of dire things would happen."

Norton had had enough. He knew that his ideas were sound, but it was quite clear that he would have to go elsewhere to have the chance to put them into practice. He went to Mr. Viall to ask for a few days off to go to look for someone who would give him his chance. Viall was as obstructionist about this as he had been about all of Norton's ideas, but Norton went anyway. In only a day Norton was back and told Viall that he had found someone. Viall asked who it was, and when Norton told him, retorted: "If the Norton Emery Wheel Company are going to make grinding machines, we will give them away." Little did Viall realize that he was giving away right then and there the talents and genius of Charles H. Norton. But this was a day when industrial leaders did not recognize the economic value of the technological genius, such as Charles Steinmetz and Charles Kettering, to name but two well known to the general public.

Triumph of Heavy Production Grinding— The Norton Company

At the Norton Company[6] Charles H. Norton found the encouragement and support he needed, especially in the per-

5. This idea Norton later showed to be untrue, and in any case the grinding wheel was an insignificant fraction of the total cost. Norton's conception of the economic saving which results from more powerful machine tools operated at capacity, has only in recent years been recognized in lathe and milling machine practice.

6. The name of the Norton Emery Wheel Company (since 1906 the Norton Company) derived from the man who founded the original pottery

son of Charles L. Allen, one of the original officers of the Norton Emery Wheel Company, who had helped Norton with some of his grinding wheel problems at Brown & Sharpe.

Beginning in March, 1900, Norton designed the 18-inch by 96-inch machine, with wheels 24 inches in diameter and 2 inches in width. This machine embodied Norton's ideas of the features required in a *precision production grinding machine*: heavy and rigid construction, high power available to drive the much larger and wider grinding wheel, and the use of precision cross feed to give the precision of which grinding is capable. Two of these machines were completed in November of that year (Figs. 43a and 43b). Viall came over to look at them and bluntly told Norton that his machine was all wrong. Norton dryly records: "Fortunately he did not say that to the Directors. He said it to me, and naturally I did not repeat it to the Directors." In 1925 Norton received the John Scott Medal for the design of this revolutionary machine.

The first machine was sold to the R. Hoe Company, manufacturers of printing presses in New York, who used it in continual operation for nearly thirty years before, in 1928, it was again in the possession of the Norton Company. They later presented it to the Ford Museum, where it is now on display. Although it was not always easy to overcome the inertia in machine manufacture and get the Norton grinding machines introduced, the editors of the *American Machinist* recognized the importance of Norton's ideas almost immediately,[7] and so did the automobile industry. The *American Machinist* noted the heavy proportions, the powerful drive, and the planning for economical production. The heavy bed, ribbed and trussed inside, had a three-point

firm in 1858, Franklin Blackmer Norton, no kin to Charles H. Norton. F. B. Norton's association with the company was only very brief during the years when they made grinding wheels. When Charles H. Norton came to Worcester it was as Chief Engineer of the separate, newly-formed Norton Grinding Company. In 1919 the two companies were "married"; they retain their identities but have a joint pocketbook and are managed as one.

7. *American Machinist,* 1901, p. 360; 1902, p. 1584.

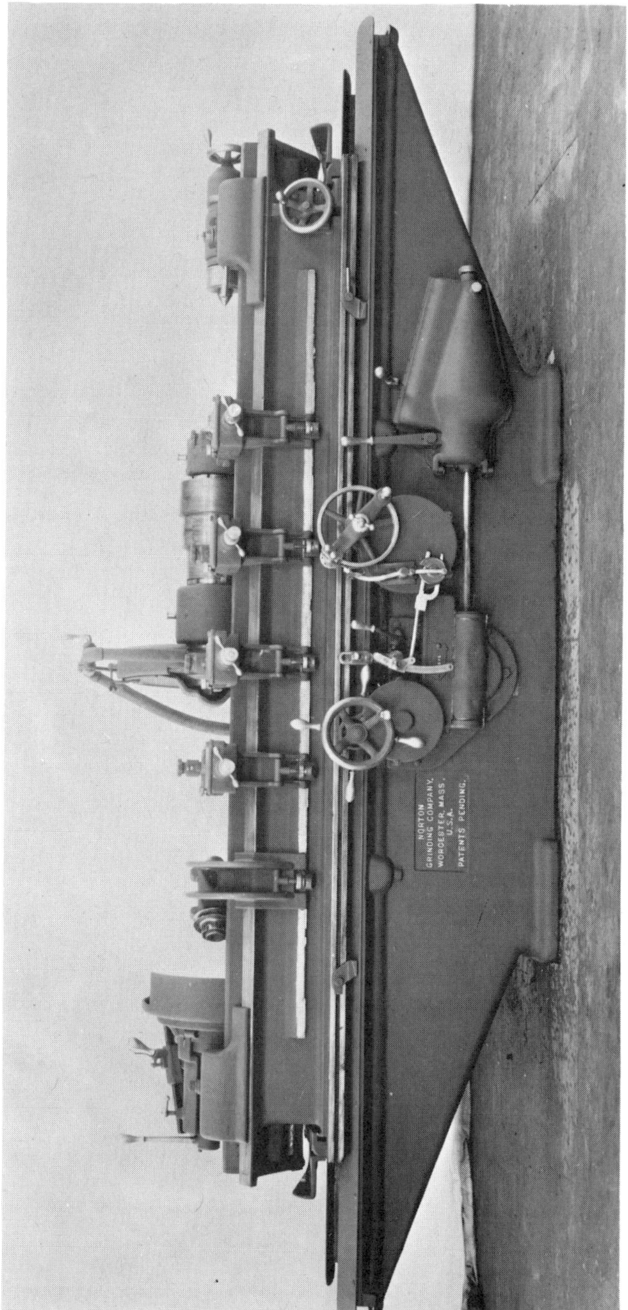

FIG. 43A NORTON'S ORIGINAL HEAVY PRODUCTION GRINDING MACHINE, 1900 (Front) (*Norton Company*)

FIG. 43B NORTON'S ORIGINAL HEAVY PRODUCTION GRINDING MACHINE, 1900 (Rear) (*Norton Company*)

FIG. 44 MODERN AUTOMATED CRANKPIN GRINDER (*Norton Company*)

suspension to prevent distortion, one point directly under the wheel, the other two at the front under the ways. To prevent any sagging between these two points they were not put at the ends of the bed, which were instead supported by two heavy knees. A wide range of speeds was available for the speed requirements of various types of grinding wheels and work. The machine had very rapid traverse of the work, one width of the wheel per turn of the work, in order to have the full width of the wheel continuously grinding, which removed the maximum of metal with a minimum truing of the wheel. Both longitudinal and cross feeds were automatic, the cross feed being high precision. An integral part of this machine was a large tank and centrifugal pump for the coolant.[8] Fifty gallons per minute were available to prevent distortion of the work from changes in temperature at the high rates of metal removal.

Success in Actual Production

Norton turned his experiments to the grinding of automobile crankshafts in 1903. As the result of his work it was found that the Norton grinder could do in fifteen minutes what had previously required five hours of turning, filing, and polishing. This process was one of the most difficult precision operations in making an automobile engine, for relatively large amounts of stock had to be removed from the crankpins, bearing bodies, and sidewalls. Dimensional accuracy and surface finish standards were also demanding. Norton's plunge grinding was the answer.[9]

By 1905 the Norton Company was grinding large numbers of automobile crankshafts on a production basis, and using plunge grinding to grind the entire width of the pin, as well as the fillets, with a single wheel. At this time the automobile manufacturers did not have their own grinding machines, because a grinding machine operator demanded

8. This feature had been patented by E. R. Hyde and A. Valentine, No. 375,821 of Jan. 3, 1888.
9. For details of actual crankshafts and the time required see Colvin and Stanley, *American Machinist Grinding Book,* N. Y., 1908, Ch. XII.

35 cents an hour, whereas lathe operators got only 30 cents an hour. The crankshafts for practically all American automobiles were shipped at this time to the Crankshaft Department of the Norton Company for grinding and balancing.[10]

In 1905 the Locomobile Company was the first to purchase a machine for grinding their own crankshafts. They were soon followed by Pope Hartford, Thomas Flyer, Haynes, and Clark Motors. Finally Ford ordered thirty-five machines, and grinding was a definitely established production method in the automobile industry. As the automobile came to be mass produced this operation required speed in order to meet cost and production requirements. From this beginning came the highly specialized, fully automated crankpin grinder shown in Figure 44.

Another problem of the automobile manufacturers was met in 1910. At that time the hardened camshafts which operate the valves in the automobile engine could only be made by mounting the separate cams on a shaft and grinding them on plain cylindrical grinders fitted with hand-operated cam-grinding attachments. The completed cams were then secured on the cam shaft in as nearly the correct position as possible. This was a long and expensive process to get a not very accurate or reliable automobile camshaft. Precision camshafts made in one piece of hardened alloy steel were made possible because grinding-machine engineers made repeated visits to motor-car plants to study their practices and their needs. Landis[11] patented his camshaft grinder in 1912. This provided for automatic feed from one cam to the next along the camshaft. A pattern shaft was geared to the work and carried master cams larger than the work, to reduce motion and thereby increase accuracy. Out of this work also came the Norton integral cam grinding attachment and later the automatic indexing arrangement.

10. *Grits and Grinds,* July 1910.
11. See A. B. Landis patent No. 1,017,879 of Feb. 20, 1912, Also H. T. Shearer patent No. 1,115,596 of Nov. 3, 1914, and his patent No. 1,156,323 of Oct. 12, 1915, for an attachment for grinding camshafts on a plain or on a universal grinder.

These devices were but the first of many of Charles H. Norton's machines which made possible the manufacture of better automobiles, more rapidly and at less cost by the use of production grinding.

The early automobile industry gave Norton a chance to demonstrate in actual production that his grinding machines and his technique of grinding not only permitted mechanical improvements in the automobile not otherwise possible, not only produced a better product than previous methods, but that all this could be done more rapidly and at less cost.

All of this advance required also a better understanding of the action of the grinding wheel in cutting metal. In 1898 the Norton Company had established a tiny research laboratory, called a Department of Tests, with George Jeppson in charge. This was the first laboratory for research on abrasives. By 1912 the research laboratory had four departments—mechanical, chemical, ceramic, and organic—and twenty-one employees. Through the facilities and cooperation of the abrasive engineers of the Norton Company, wheels were produced which, properly used, *cut* the metal into the same kind of curled chips as were found in using other metal-cutting tools. These chips were, to be sure, extremely small, but since millions of them are cut per minute, it is possible in many cases to remove more metal in less time than with other metal-cutting tools. In fact the original Norton machine of 1900, using a grinding wheel four times larger than previous machines, cut sixteen times faster. It cut one cubic inch of steel (over ¼ pound) per minute, working to an accuracy of .001 of an inch. In short, grinding is not a slow metal-cutting process, but is actually one of the most rapid.

What Norton had done was to demonstrate conclusively that grinding was a rapid, flexible, and economical means of production.

Charles H. Norton, while the stoutest defender of ideas which he knew to be right, was also a man to give full credit where it was due. In his letter of December 6, 1929, to Howard Dunbar, of the Norton Company, he gave the Norton Company itself full credit for its share. "Who was it that furnished the capital with which to spread the knowledge of my experiments and enable me actually to force on the world . . . the methods and machines that have led . . . . [others] to adopt those methods and to build machines upon those principles?" He might have added that the Norton Company not only put up the capital and took the risk, but they gave him their confidence and assistance in his work, and the association of men vitally interested in the grinding machine as a production tool.

We should also not fail to note the part played by the new automotive industry and the men of vision creating it who were prepared to accept new methods which would enable them to give the public a better automobile at a cheaper price. They provided the economic incentive, which was as necessary to Norton's work as his revolution in machine tools was to the revolution they were bringing about in the American economy and way of life. Only through the coming together of all these factors can we account for this dramatic change in production from 1900 to 1915.

SPECIALIZED GRINDING MACHINES MEET THE DEMANDS OF INDUSTRY 1885 to 1930

The Bicycle and Ball Bearings

We have already seen that specialized machines for grinding lenses and other optical parts, for grinding precious stones, and for making needles and mirrors appeared quite early. At Brown & Sharpe, Charles H. Norton had designed a special grinding machine for the triple cylinders of the air brake as early as 1888. In 1899 Pratt & Whitney were making precision lathes and milling machines for watchmakers. In order to get the high precision required for this work they built machines specially designed to use diamond laps for grinding the bearings and spindles for the headstock of these tools.[1] All these grinding machines, however, arose to meet the peculiar requirements of a special product being manufactured in rather limited quantities.

Joseph R. Brown had used grinding to meet the needs of the economically important manufacture of sewing machines by 1860, but most of these grinding machines were not what one could really call specialized. With the use of a few fixtures a general-purpose grinding machine was quite adequate. Until somewhat later neither the technical problems involved nor the volume of sewing-machine production required specialized grinding machines.

The story, was, however, quite different with the application of grinding to the bicycle, the steam locomotive, and the automobile.

Ball bearings were used on bicycles as early as 1877, but they proved to be not very satisfactory until the 1890's,

1. *American Machinist,* 1899, p. 31.

and would have been less so had they been tried on machines of heavier loads. The successful ball bearing depends upon having the balls themselves perfectly spherical and all of identical diameter. The ball must run in races perfectly circular, perfectly concentric, and of exact dimensions. Not only must the balls and their races be machined to a fine surface finish, but all these dimensions must be held to close tolerances, and all these parts must be hardened. Only grinding could deal with this problem.

The production of hardened balls had been accomplished by Barker and Holt in 1853, but their balls did not meet the standards necessary for ball bearings. It was not until the work of Henry Richardson in 1887 that it was possible to produce *precision* hardened balls. Richardson's ball bearing grinder[2] used essentially the principle of Barker and Holt's, but his machine incorporated the features necessary to produce precision work. From his invention the ball bearing became possible.[3]

Finish, exact dimensions, and precise circularity and concentricity of the races for ball bearings were provided for by the specialized grinding machine built in 1896 by the Diamond Machine Company, of Providence, Rhode Island. On this machine two grinding heads were provided in order to do both ends of the bearing at once and thus to ensure that the ends were perfectly true with each other. This machine also had a number of features designed to simplify and speed up the work.[4]

The application of the ball bearing to the bicycle had a special problem. The bicycle wheels had their spokes attached to a shell containing one half of the ball bearing race, called the cup. The other half of the race, called the cone, was secured to the shaft about which the wheel rotated. This shaft was held by nuts to the frame of the bicycle (Fig. 45). It was, of course, necessary that the hardened balls roll on a hardened cup and cone. Since both of these pieces

2. Patent No. 365,407 of June 28, 1887.
3. *Abrasive Industry,* 1926, p. 163.
4. *American Machinist,* 1896, pp. 132 and 847.

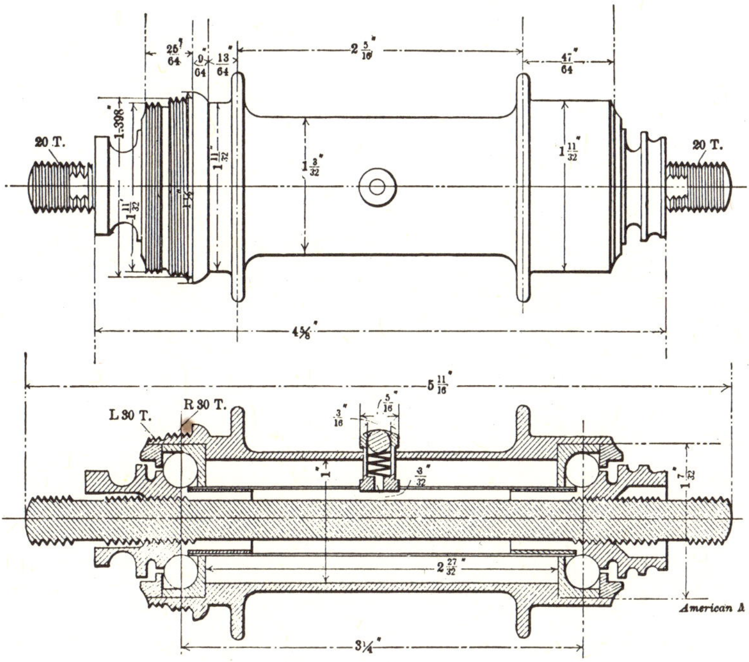

FIG. 45 BICYCLE BALL BEARINGS, 1896 (*American Machinist*)

were quite light and of irregular shape, it was impossible to harden them without distortion. Grinding was the solution.[5] The machine which Brown & Sharpe had designed for grinding sewing-machine needle bars was adapted by some firms to the bicycle's cups and cones. The Brown & Sharpe machine had the advantage of having one work-carrying spindle and two grinding-wheel spindles. In this way both the interior and the exterior could be ground at the same time. This saved labor, but, more important, it ensured concentricity.

The Lozier company, however, built their own grinding machines for ball bearings prior to 1896. Since this firm had been making sewing machines before they got into the

5. *American Machinist,* 1896, p. 517.

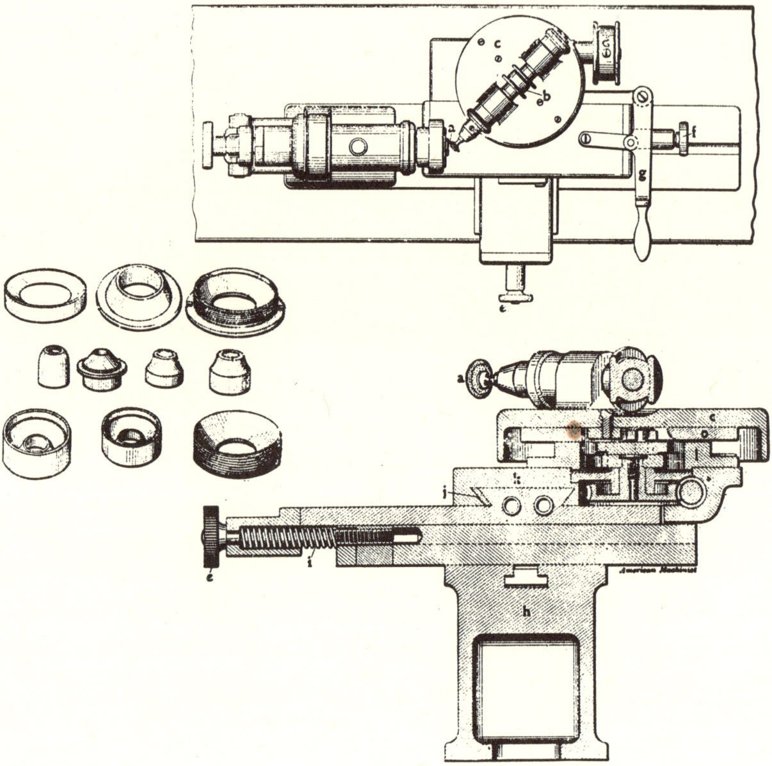

Fig. 46 Pratt & Whitney Ball Bearing Grinder, 1899
(*American Machinist*)

bicycle business, the unknown designer of their special machines probably was familiar with the Brown & Sharpe machine. However, concentricity and precision were carried in the Lozier machines to the limit; their own spindles were all mounted in ball bearings. The shape of the ball race was made by a formed wheel constantly checked by a gage. The completed parts were carefully tested against standard cups, cones, and balls to ensure accuracy up to .00025 of an inch.

The grinding machine for making the cups and cones for bicycle ball bearings became more or less standardized in 1899 when Pratt & Whitney brought out the machine[6]

6. *American Machinist,* 1901, p. 737.

FIG. 47 PRATT & WHITNEY GUN MOUNT GRINDER, 1906
(*American Machinist*)

shown in Figure 46. An emery wheel was so arranged that by a system of cams its movements could be adjusted to turn out the profiles of the races used by several different makers of ball bearings. The grinding of the cups required very small grinding wheels, and in order to attain the necessary surface speed, provision was made to give the grinding wheel as much as 28,500 revolutions per minute. This speed required ball bearings specially designed for use in this machine, but they could run for hours without overheating or showing any appreciable wear. A special tape belt drive was also required by this high speed. A collet chuck operated by a treadle permitted high production rates.

Important as all this development was to the bicycle industry, it quickly had wider applications in other industries. A specialized use of the technique was used to produce the four-point bearings of the Auburn Ball Bearing Company by 1901.[7]

In 1906 Pratt & Whitney built a much larger bearing grinding machine[8] to grind the large ball races and rings of gun mounts (Fig. 47). This machine could work parts from 12-inch diameter inside to 40-inch outside. It had three grinding heads. The one at the right ground the channel or race. By means of a templet the motion of this grinding wheel was controlled to produce any curve desired for the cross section of the race. The grinding head at the left ground the face of the ring, and the one on the column ground the outside and inside of the ring.

The grinding machine not only made the bicycle possible; in so doing it drew attention to the possibilities for grinding in similar problems on a larger scale. But grinding and the bicycle did something of great importance for all machinery: they developed the techniques by which ball bearings (and later roller bearings) led the attack on those two enemies of all machinery—wear and friction.

The Railroad and Heavy Grinding

The steam locomotive was certainly not made possible by the use of grinding machines, but it did have a number of technical problems that were solved by grinding. Grinding improved the locomotive and railroad cars by making higher speeds possible with less wear and vibration and with maintenance costs substantially reduced.

Special machines for grinding the link bars of locomotives were in use in England as early as 1887. Beyer, Peacock & Company brought one out in that year, of the vertical spindle type.

In 1891 the chilled wheels of cars were ground true and smooth.[9] This process gave much longer wear of the wheels and their journal bearings, for pounding and vibration were reduced. Wheels which had been given flats by

7. *American Machinist,* 1901, p. 737.
8. *American Machinist,* 1906, p. 703.
9. *American Machinist,* Jan. 8, 1891, p. 1.

Fig. 48 Springfield Car-Wheel Grinder, 1891 (*American Machinist*)

slipping or by jammed brakes could be restored to use at little cost on the machine shown in Figure 48. The heavy construction of this machine was required to carry the load of two wheels and their axle. Although the machine could grind the slight tapers desired on the wheel and its flanges, it was not a precision process; all that was required was a smooth finish and good circularity of the hardened wheels. The reader will note the use of a large grinding wheel for this purpose, as well as a tank and centrifugal pump to supply plenty of coolant for this heavy grinding. The Pennsylvania Railroad found that this machine repaid the trouble,

Fig. 49 Springfield Journal Bearing Grinder, 1891
(*American Machinist*)

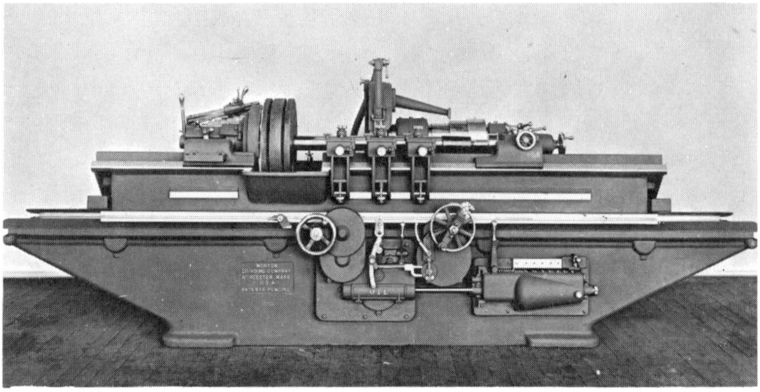

FIG. 50 NORTON'S GAP GRINDING MACHINE FOR LOCOMOTIVE PISTON RODS, 1902 (*Norton Company*)

especially for passenger service. In ten hours it could finish six to eight new wheels, or refinish five to seven old wheels, at a total cost of 28 cents per wheel.

The Springfield Emery Wheel Manufacturing Company of Bridgeport, Connecticut, also made another specialized grinding machine for grinding "car boxes" or journal bearings.[10] This machine (Fig. 49) was also double-ended and had automatic feed along the axis of the bearing. Its grinding head is of some interest, for it was a kind of chuck that held three emery plugs forced out against the work by cams controlled by a screw mechanism.

A more highly developed specialized machine for locomotive grinding work was the gap-grinding machine for piston rods designed by Charles H. Norton in 1902 (Fig. 50). Norton's machine could do the original grinding of the piston rods, as well as grind slide-valve stems with yokes. An elaborate set of steady-rests was provided in order to assure precision along the length of the piston rods, which could be ground up to an overall length of 8 feet.[11] This machine could also regrind piston rods up to 4 inches in diameter and with their pistons, up to 28 inches in diameter, still in place. By so doing it was possible to increase the life of loco-

10. *American Machinist,* Apr. 16, 1891, p. 11.
11. *American Machinist,* 1902, p. 1629.

Fig. 51 Vertical Planetary Grinder for Locomotive Links, 1914 (*Halsey*)

motive piston rods and at the same time improve their quality, both with less expense.

The application of grinding methods to the manufacture and maintenance of railway rolling stock was also developing rapidly in England and Germany; in fact, both countries ground locomotive cylinders for overhaul some years before the practice was common in the United States. Mayer and Schmidt, for example, put on the market in 1909 a planetary internal cylinder grinder for steam and gas engine cylinders.[12] This machine was also supplied with an attachment for grinding the arcs of locomotive reversing links.[13] Friedrich Schmaltz developed a vertical planetary grinding machine by 1914, especially for locomotive work (Fig. 51). The advantages of a horizontal table were considerable in handling heavy work. The planetary head could be used for both internal and external

12. *Machinery*, Jan. 1909, p. 354. For the principle of the planetary grinder see p. 131, Fig. 60.
13. B. E. Parks had patented a special grinding machine for this purpose. U. S. Patent No. 339,010 of Mar. 30, 1886.

FIG. 52 PLANETARY GRINDER FOR LOCOMOTIVE CRANK PINS, 1914 (*Halsey*)

grinding, and a supplementary table shown in our illustration permitted grinding locomotive links with their eccentric rod pins in place. Not only the link arc and its holes, but also the eccentric rod pins, could be ground at the same setting. This firm also made a planetary grinding machine of the horizontal type which could grind locomotive crank pins after they had been forced into place on the driving wheels and their shaft[14] (Fig. 52).

By 1915 grinding—cylindrical, internal, and surface—was used in locomotive work as a matter of course (Figs. 53 and 54). Charles H. Norton said later, "the only markets for grinding machines in those days were the railroads and the general machine shops."[15] Grinding reduced cost and increased output, especially in the repair of worn parts.[16] Specialized grinding machines developed to meet the needs of locomotive shops were on the open market in 1925 manufactured by Norton, Gisholt, and others.[17]

Specialized heavy grinding machines were clearly making their contribution to the transportation of the world's goods.

14. F. A. Halsey, *Methods of Machine Shop Work*, N.Y., 1914, p. 276.
15. Norton Company, *Salute! Mr. Norton,* Worcester, Mass., 1941.
16. *Abrasive Industry,* Apr. 1925, p. 116; May 1925, p. 141.
17. *Abrasive Industry,* Mar. 1925, p. 96; Apr. 1925, p. 126; May 1926, pp. 136-152.

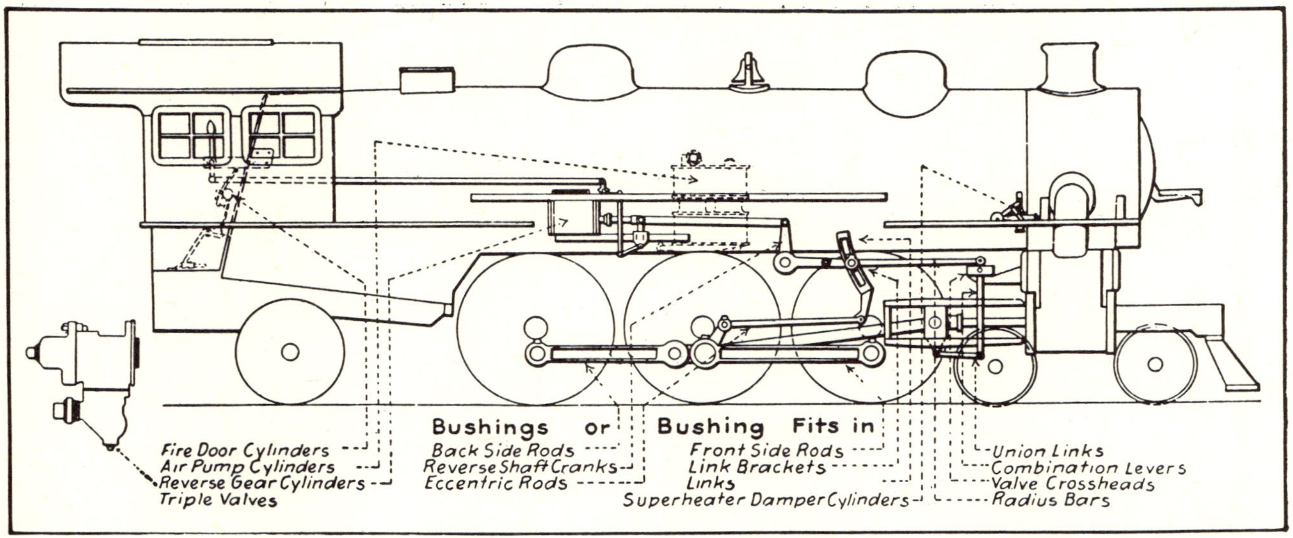

FIG. 53 LOCOMOTIVE PARTS USING SURFACE GRINDING (*Abrasive Industry*)

RAILROAD EQUIPMENT PARTS FINISHED BY SURFACE GRINDING	
Guide bars	Straps for main rod
Slide valves	Simplified steam chest steam-pipe casings
False valve seats	Covers for steam chest
Flange connections for triple valve	Pedestals for truck wheels
Links	Castings for joint ring for metallic packing
Joints for exhaust pipe	Rocker shaft split bushings
Eccentric straps	Center castings
Guides for valves	Filling in plates for guides
Nozzle stands	Plates and covers for bypass valves
Expansion rods	Steampipe split castings for air tight casings
Feed water heater flat joints	Pedestal jaws for engine trucks
Finishing crank arms	Locomotive whistle flange
Cellars for driving box	Crossovers
Binders	Crosshead keys
Trailer truck boxes	Reverse shaft split bushings
Switches	Split pulleys
Frogs	Shears and knives
Shoes and wedges for driving box	
Brasses for main rod	

FIG. 54 LOCOMOTIVE PARTS USING INTERNAL GRINDING (*Abrasive Industry*)

The Automobile and High Production

A volume should be written about machine tools and the automobile industry. In no industry has the technical development of tools been so crucial to all its aspects, and no industry has had such a great effect on the far smaller machine-tool industry. Not only did the automobile industry after 1900 become the largest single customer of the machine-tool industry, taking 25 to 30 per cent of the output, but it actually increased its volume consistently.[18] After 1930 the

18. Article "The Influence of the Automobile on Machine Tools," in *Mechanical Engineering,* Aug. 1921, p. 529. Also the series of articles "The Automotive Industry and the Machine Tool Builder," beginning in *Machinery,* Oct. 1924.

aircraft industry began to take about an equal percentage of machine-tool production. Because the automobile manufacturers very early utilized the techniques of mass production to obtain large volume, they required and could afford more extensive use of highly specialized machine tools. These forces are still operating at the present time. The Ford Motor Company's engine plant at Cleveland, Ohio, is in a very real sense one enormous specialized machine tool intended to take raw metal, cast it, and machine it into finished engine blocks and heads, by means of fully automatic and automated machinery.[19]

In all this interaction between the automobile and machine tools, a central place was held by the grinding machine, partly because it could provide higher accuracy and better finish,[20] but principally because it permitted the automobile designers to use the advantages of the new alloy steels. Their strength, and corresponding lightness of parts, presented difficult problems if machined before heat treatment, and insoluble difficulties when hardened, unless the techniques of grinding were employed.

It has been estimated that over three hundred separate grinding operations are performed on the parts of a modern car. While the early cars were simpler and made with less use of grinding, they also had many different parts which required this process, a large portion of which were made on grinding machines designed especially for that operation. Even to try to list them all would not serve our purpose here, but a number might be mentioned simply to show their variety, before we go on to consider those grinding operations, which were crucial and which illustrate the special contributions which grinding made to the automoble industry.

In 1913 the Ford plant had thirty-three special-purpose machines in operation grinding the vanadium steel drop

19. Even if the "transfer" from the automated foundry to the automated machining plant is, at the moment, of the crudest sort—men and carts! Human muscles also load the castings onto the first stage of the automated machinery!

20. For the use of grinding in the automobile and other industries by 1922, see Fred B. Jacobs, *Production Grinding*, Cleveland, Ohio, 1922.

forgings for their transmission shafts. These machines, designed and made by the Norton Company, used a large, wide wheel, with a 20-inch diameter and 6-inch face, and with a surface speed of 6000 feet per minute to grind the shaft rotating at 150 revolutions per minute. The work was done in a single plunge cut, to tolerances of .0005 of an inch. The aluminum oxide wheel was trued about every 65 shafts, and production was at the rate of 550 shafts every eight hours.[21]

By 1925 grinding was being used for producing hardened and tough steering knuckles on a special machine using two wheels grinding at once. Axle parts were also made on special double-wheel grinding machines using wide wheels and a plunge cut.[22]

The "Silent Knight" engine was only made possible by specialized grinding machines to produce the close fits required on the hardened, thin cylinders which made up its sleeve valves.[23] The close fits on the more common type of poppet valves used in automobile engines also required grinding of their seats and their stems for proper action and long life.

The increasing size and speed of the automobile and the demands for safety from the increasing highway traffic, led to the introduction of four-wheel brakes, which in turn put a new emphasis on adequate design and manufacture of brakes. In 1925 Landis put out a special machine for grinding brake drums.[24]

Very early in the development of the automobile the demand for quiet gears hardened to reduce wear and the effects of careless gear shifting, and the requirements of gears for the differential drive, led to ground gears.[25]

21. Advertisement of the Norton Company, in *Machinery,* Jan. 1915, p. 120. Also reproduced an *Abrasive Industry,* Apr. 1925, p. 6.
22. *Abrasive Industry,* Jan. 1925. For later development of simultaneous grinding by several wheels, see *Grits and Grinds,* Dec. 1957, p. 3.
23. *Abrasive Industry,* Mar. 1925, p. 73 and Apr. 1925, p. 111.
24. *Abrasive Industry,* May 1925, p. 6.
25. *Abrasive Industry,* Aug. 1925, p. 246 and J. J. Guest, *Grinding Machinery,* London, 1915, pp. 361-372. Also see R. S. Woodbury, *History of the Gear-Cutting Machine,* The Technology Press, Cambridge, Mass., 1958, pp. 121-124.

The widening use of grinding was by no means restricted to Ford or even to the United States. Olds used grinding extensively.[26] Although grinding techniques came a little later in the English automotive industry than in the United States, it was in full use by Vauxhall in 1925.[27]

In the business slump of the early twenties, reflected as usual in the "feast or famine" machine-tool business, the grinding industry saved itself from threatened bankruptcy by turning out machines especially designed for rebuilding automobile motors by regrinding.[28] The larger garages purchased machines for regrinding automobile cylinders, and thousands were sold to enable the man who could not afford a new car to make the old one do, by the help of his local garage mechanic.

Grinding even entered into the body and appearance of the automobile by 1925. The sheet metal of which the body was made had to be rolled in a mill. In early cars it was so imperfect that the defects showed through even the many coats of paint which were put on, baked, and then rubbed down. Chilled rollers ground to a mirror-smooth finish permitted manufacture of sheet steel for automobile bodies free of these defects. The Landis Tool Company had been the leader in this development since 1905.[29] One of their early machines for grinding the rolls for sheet steel mills is shown in Figure 55. And grinding quickly smoothed up the large dies on which these body parts were stamped into the form desired.

Grinding was also frequently applied with specialized machines to clutch housings, connecting rods, thrust washers, intake manifolds, axle housings, and valve seats.

In 1905 the Landis Tool Company put on the market a specialized automobile crankshaft grinder[30] (Fig. 56). In

26. *Abrasive Industry,* July 1925, p. 198 and Aug. 1925, p. 233.
27. *Abrasive Industry,* July 1925, p. 210.
28. *Abrasive Industry,* Jan. 1925, p. 6; Apr. 1925, p. 106; Aug. 1925, p. 240; Sept. 1925, p. 281; Nov. 1925, p. 354.
29. See their long series of patents for improvements in roll grinders beginning with A. B. Landis Patents No. 785,258 of Mar. 21, 1905, and No. 861,738 of July 30, 1907.
30. *American Machinist,* 1905, p. 547 and *Abrasive Industry,* Nov. 1925, p. 7.

order to eliminate torsion in the crankshaft while grinding, the work was held by its journal bearings in clamps mounted offset on the two live heads. These heads also carried counterbalances to offset the unbalance of the crankshaft when grinding its pins. A steady rest was provided to give extra support to the pin being ground, and a special truing device was mounted on the top of the steady rest in such a way that the wheel could be trued without removing the work from the machine. The diamond-point truing device could form the arcs needed on the wheel edges to grind the fillets required on the crankshaft by the bushings which keep the connecting rods in alignment. The radius of these arcs was adjustable up to ½ inch.

In connection with the work of Charles H. Norton we have already traced the introduction of grinding techniques into the making of automobile crankshafts and camshafts and seen how grinding actually made these important parts of the engine possible, how it gave them high precision and at reduced time and cost. Similar results were also to be found in grinding the parts which form the heart of the gasoline engine: the cylinder, the piston, and the piston rings which retain the burning gases and force them to do work on the crankshaft by means of the connecting rods.

The piston-ring problem had already been met in the gas and steam engines. It was essential to grind their top and bottom surfaces in order to get a tight joint with the piston, and it was necessary to grind their outer circumfer-

FIG. 56 LANDIS AUTOMOBILE CRANKSHAFT GRINDER, 1906
(*Landis Tool Company*)

ence in order to get a tight fit with the cylinder wall. For the slower action and lower pressures of these earlier engines these problems were not difficult, but with the higher speeds, temperatures, and pressures of the automobile engine they assumed a much greater importance.

In 1902 Reinecker, in Germany, put out a special surface grinder for grinding the top and bottom of piston rings for steam and gas engines. It had a grinding wheel on horizontal axis and a magnetic chuck mounted on a rotating table.

The Heald Machine Company announced in 1904 their 6-inch rotary grinder, designed by James N. Heald especially for grinding the sides of piston rings for the rapidly growing automobile industry. With this machine we have the appearance of the other principal in what might be called "The Drama of Barber's Crossing," for in that section just outside Worcester, Massachusetts, James N. Heald and Charles H. Norton were to develop the grinding methods and machines which made the American automobile industry possible. The

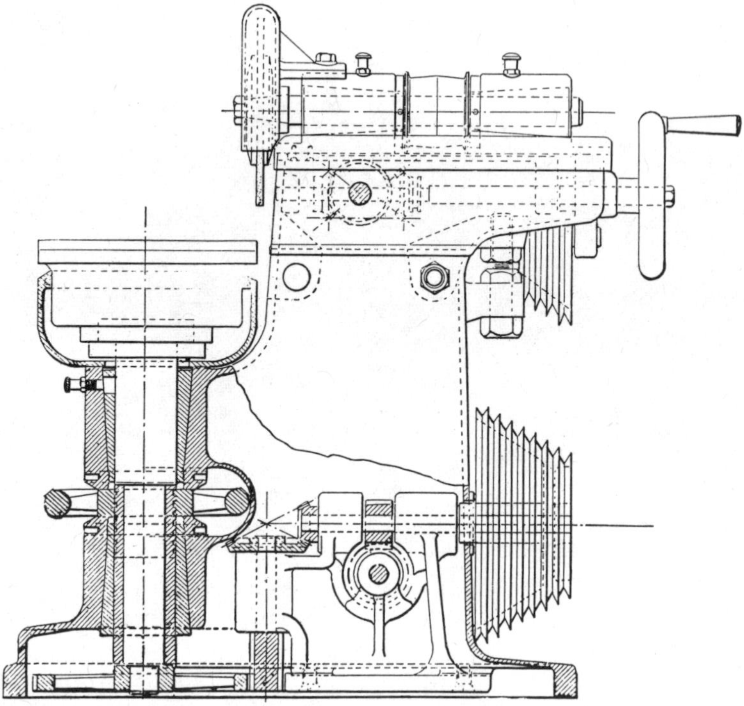

FIG. 57 HEALD PISTON RING GRINDER, 1904 (*American Machinist*)

two firms with which they were associated now stand just across the street from each other—the Norton Company and The Heald Machine Company.

Heald's first piston-ring grinder had a wheel mounted with its axis horizontal, and a magnetic chuck[31] on a rotating table, but a micrometer adjustment was provided to permit precise dimensions between the two faces of the piston ring (Fig. 57). It was pointed out that by precision grinding its piston rings, the automobile engine would deliver more

31. The invention and development of this device will be described in a later monograph on the History of Jigs, Fixtures, Arbors, and Chucks. The magnetic chuck was especially important for surface grinding because it made possible grinding thin parts which would be distorted by the usual means of holding the work. It was the invention of O. S. Walker in 1896, another Worcester man like Norton and Heald.

power. The machine was also useful for grinding dies, disks, cutters, and thrust collars.[32]

In 1906 a specialized grinder for doing the outer circumference of piston rings had been developed. The ring was put on a form intended to hold it in the shape it would have in use. This form was then slowly rotated on a vertical spindle as the circumference of the piston ring came up against the edge of a cup grinding wheel. A better method was described by Colvin and Stanley in 1908, in which a number of rings were slid onto a special mandrel and had their outer circumference ground at the same time by a conventional cylindrical grinding machine.

Although the cylinders of steam and gas engines had already been ground for some time, there were special problems in the manufacture of cylinders for automobile engines. In the first place, they were much smaller, and this involved problems of the size of the grinding wheel and its speed. Next, because they were smaller and because of the need for rapid dissipation of heat from the high temperatures produced in them, automobile cylinders had to have rather thin walls. Further, internal combustion engines had to be light yet strong; they therefore were made of a hard, close-grained iron which was always difficult to machine accurately. And last, the smaller size of the gasoline engine meant that it could advantageously be built with several of its cylinders in one cylinder block.

Through his contacts with the automobile builders, James N. Heald recognized before 1905 that engine performance was handicapped by rough and inaccurate cylinder bores. The practice at that time, and in fact with some automobile engine builders for some years afterward,[33] was to bore out each cylinder, ream it, and lap it for finish, each cylinder being done separately. Heald recognized very quickly the troubles with this procedure, and found the solution in grinding these cylinders.[34]

32. *American Machinist,* 1904, p. 433.
33. See *American Machinist,* 1917, p. 237 for the techniques employed in making the Wiedely 12-cylinder engine.
34. James N. Heald, "Grinding Automobile Cylinders," in *American Machinist,* 1910, p. 252.

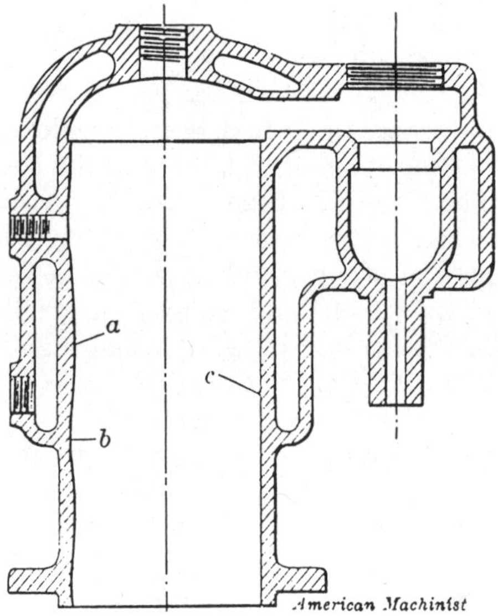

FIG. 58 EFFECT OF BORING AUTOMOBILE CYLINDERS, 1905
(*American Machinist*)

The walls of gasoline engine cylinders are thin. They will therefore spring away from the boring tool, especially if the tool strikes a hard spot in the iron casting. The result is shown at "*a*" in Figure 58. A soft spot will allow the tool to cut too deeply, as at "*b*." The resulting high spots will chafe against the piston rings; the low spots will allow the gases to leak by the piston ring with consequent loss of power. For cylinders having port holes the results are even worse, for the boring tool jumps into and out of these openings and breaks off their corners. Automobile manufacturers tried to correct these defects by reaming and found that reamers gave similar effects, which were made worse in cylinders having ports in one side, for the reamer tended toward that side and the cylinder was then thrown out of round. In addition, it took much careful sharpening to maintain the exact size of the reamers, and such large reamers were costly. Heald pointed out that the practice of lapping the cylinder walls merely produced a fine finish but did not ensure either a round or a straight hole. Heald was sure that the solution was to grind the cylinder holes.

Some manufacturers claimed that grinding would leave the metal of the cylinders charged with some abrasive, which would lead to wear. Heald pointed out that grinding was being used on the spindles of the highest precision machine tools and other machinery, and that in any case, careful chemical analysis had failed to reveal any abrasive left on any ground cylinder surface. Had Heald extended chemical analysis to the results of their practice of lapping cylinders, he would almost certainly have found abrasive in their walls.

Other automobile manufacturers objected that in a ground cylinder it would take too long for the piston rings to "come to bearing." Heald said that this was the case only if the ring itself had not been ground on its outer circumference. Some engine makers had even put ground piston rings into cylinders merely bored and reamed, which of course only spoiled the ground finish on the piston ring. Heald's idea was to grind both the piston ring and the cylinder, and the engine will deliver its full power right from the factory. Later, this process was extended by lapping a fine finish on piston rings and cylinders and on crankshafts. Then a "breaking in" period was no longer required for new automobile engines.

Heald saw the solution clearly: rough bore the cylinders and then grind them, to give cylinder holes which are accurately straight and round and of the smooth finish which only grinding can produce. Precision roundness of the hole is assured by the fact that the grinding wheel itself is trued perfectly round, rotates at high speed, and is carefully balanced on a well-supported spindle. Its cutting action is such that its pressure on the work is extremely light, and it is completely unaffected in its precision by either hard or soft spots in the metal. The table travel of Heald's grinding machine was to be itself precision straight, to produce cylinders which did not deviate from straightness by more than .00025 to .0005 of an inch.

The advantages of multiple cylinders for internal combustion engines were also apparent to engine designers. But this meant a large, heavy, and awkward casting which could

FIG. 59 HEALD CYLINDER GRINDER WITH PLANETARY SPINDLE, 1905
(*Halsey*)

not possibly be rotated as other internal grinding work had been. Now Heald's problem was how to design a grinding machine that would meet all these requirements.

His first decision was to keep the cylinder block fixed on the table of the machine and to give the wheel a planetary motion (Fig. 59), that is, as the grinding wheel revolved at high speed on its spindle, the axis of the spindle itself moved more slowly in a circular path whose diameter could be accurately adjusted. Provision could easily be made to align the center of rotation of this spindle accurately with the axes of the different cylinders in the engine block, merely by a cross adjustment on the table to move from one cylinder to the next. This motion on accurate ways assured that all cylinders were exactly in line and with their axes exactly parallel. Since the motion was controlled by a micrometer screw, precision distances between the centers of the cylinders were certain. All this led to good alignment of the piston and its connecting rod with the crankshaft, which reduced wear and led to more quiet operation.

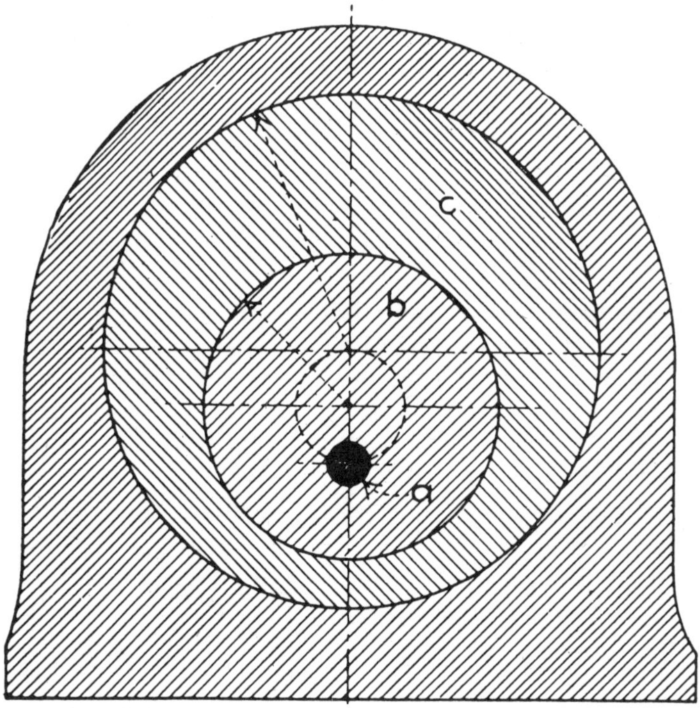

FIG. 60 PRINCIPLE OF THE PLANETARY GRINDER (*Halsey*)

The next question was how to give the grinding-wheel spindle an accurately adjustable planetary motion. This motion was achieved by use of the principle shown in Figure 60. The grinding-wheel spindle is shown at "*a*" and is carried eccentrically in a large bushing "*b*," which is in turn carried eccentrically in the greatly enlarged main spindle "*c*." This main spindle is supported in the frame as shown. These various eccentricities are proportioned so that in the extreme position, as in Figure 60, the axis of the wheel spindle will pass around the circle of maximum diameter. However, if bushing "*b*" is rotated through 180°, the axis of the wheel spindle will coincide with that of the main spindle "*c*," and the wheel spindle will not then move in a circle, but will remain fixed at the center. Therefore, by adjusting the angular position of the bushing "*b*," the center of "*a*" may be made to travel in a circle of any diameter up to a certain maximum. Its use for grinding holes of larger diameter can be extended by using larger grinding wheels.

By 1905 James N. Heald had incorporated these principles in a machine[35] so well designed that planetary cylinder grinders even today differ from it only in details, and Heald cylinder grinders were for many years standard equipment for finishing automobile cylinders, and in World War I for airplane cylinders.

The technique of using these machines was simple. The material to be ground was cast iron. It proved best to grind this dry and to remove the dust by an exhaust as shown in Fig. 59. The work was cooled by circulating water through the cylinder block, commonly at the normal operating temperature of the engine.

The results were not merely technical improvement, but production changes of economic importance as well. Because the grinder could take over after very rough boring, the expensive reaming and lapping of cylinders were eliminated, and costs were greatly reduced. Heald summed up the results in his paper in the *American Machinist* in 1910:

> The finish produced is superior to that obtained by any other process; multiple cylinders are handled as easily as single cylinders; better compression is obtained; full power is obtained at the start off; less trouble is had in regard to lubrication and

35. The first patent for a planetary cylinder grinder is that of George H. Newton, No. 791,570 of June 6, 1905. Brown & Sharpe exhibited at the Liége Fair of 1905 their #23 grinding machine based upon this patent. Similar machines were shown by Naxos-Union and Mayer & Schmidt, and a vertical type by Friedrich Schmaltz. (See *Ameriacn Machinist,* 1905, p. 351 and G. Schlesinger, *Die Werkzeugmaschinen auf der Weltausstellung in Lüttich, 1905*, Springer, Berlin, 1906.)

Planetary internal grinding machines by Mayer & Schmidt were in use in Germany and Italy for finishing automobile cylinders by 1909 (*Machinery,* 1909, p. 354). In 1914 vertical grinding machines of the planetary type were in use. (See Fig. 51). The planetary grinder for automobile cylinders has now largely been replaced by the multi-spindle honing machine.

The Heald Machine Company has on exhibit at their plant early models of many of their grinding machines mentioned in this monograph, as well as the actual 1922 development model of their internal grinder with automatic size control of 1927.

better running engines can be produced than it is possible to build in any other way.

Grinding will give you easily and cheaply parts that are interchangeable, which is so essential to high grade work and so indispensable to low cost of production. In fact the advantage of interchangeability alone would be sufficient to warrant the use of the grinding machine even if there were no other advantages to be gained.

This last paragraph may well serve as a summary of what the technical work of Charles H. Norton and James N. Heald did, not only for the automotive industry, but for all production which depends on machine tools—the development of precision production grinding.[36]

36. Lest the reader get the impression that grinding was important only to the sewing machine, the bicycle, the locomotive, and the automobile, he may refer to the application of this process to making Otis elevators and their electric motors. Grinding also made possible our skyscrapers. (*Abrasive Industry;* Nov. 1925, pp. 329-333). Many other applications are easily found described in *Abrasive Industry* and other trade journals already cited.

HIGH-SPEED PRODUCTION GRINDING

AUTOMATIC AND CENTERLESS

1905 to 1950

In the last two chapters we have seen grinding take its place in production as a widely used machining method of great flexibility, high precision, rapidity, and low cost. The grinding machine had proved to be a machine tool of great value in a wide variety of the metal-processing industries. As a result, the grinding machine became more and more automatic in the attempt to get higher production rates, grinding came to be applied to many types of work hitherto done on other machine tools, older types of grinding machines had a rebirth, and even a new type of grinding machine was invented, which soon had great importance in high-speed, light, precision grinding for production.

The grinding machine had quite definitely come out of the tool room and was now found doing all kinds of work on the production line.

Automatic Operation and Measurement

We have already seen how the application of specialized grinding machines made possible the manufacture of ball bearings, at first for bicycles and then for other machinery. Because the ball bearing brought about such an enormous reduction in friction and wear, it was very quickly applied to all kinds of machinery. The result was a tremendous demand for ball bearings of all sizes. In order to meet this demand, special automatic grinding machines were developed for high-speed production. By 1903 A. B. Landis had patented an automatic magazine feed and release for short cylindrical parts.[1] One of these made by the Norton Company and in use by 1908 is shown in Figure 61. It was designed for grinding the outside rings of ball bearing races.

1. Patent No. 734, 897 of July 28, 1903.

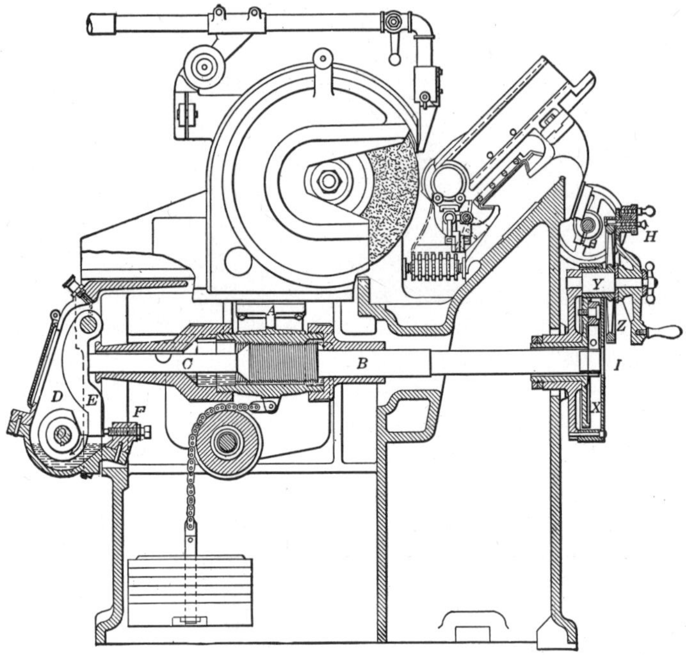

FIG. 61 NORTON AUTOMATIC GRINDING MACHINE FOR BALL BEARING RACES, 1908 (*Colvin and Stanley*)

This machine had automatic loading of the work as well as automatic control of the grinding cycle. The rings were held in a magazine feed and rolled down to be gripped in the chuck when the retaining plunger was withdrawn. After grinding, the ring dropped out of the chuck and fell on a small chain-belt conveyor and was carried to a box near the machine.

The chucking mechanism, traverse of the work across the face of the grinding wheel, and feed of the heavy grinding head to and from the work were all automatic. Truing of the wheel and sizing the work were not automatic; however, an ingenious device did make it simple to compensate for loss of wheel diameter. Using a large wheel with a wide face, little truing was required; and the ball races were easily ground to an accuracy of .00025 of an inch.[2]

2. Colvin and Stanley, *American Machinist Grinding Book*, New York, 1908, p. 33.

In 1912 A. B. Landis patented two types ot automatic feeds. One was designed for grinding the heads and ends of crankshaft pins, with the work held and fed mechanically between two opposed in-line disk grinding wheels. The other used mechanical feed and a single wheel for grinding the body of these pins.[3]

Another approach to attaining high-speed production grinding was an automatic grinding machine using several grinding wheels in succession or simultaneously, essentially to apply the principle of the automatic screw machine[4] to grinding. This type of grinding machine became the specialty of the Bryant Chucking Grinder Company of Springfield, Vermont. The one shown in Figure 62 had three independently driven grinding wheels, each having separate horizontal feed. Each of the three spindles had independent means of adjustment, the center spindle usually being assigned to the internal grinding operation. There were controlling cams and stops for each spindle. The work was held in a chuck. One advantage of this machine was that it allowed the use of the right wheel and the right speed for each operation, a common setup being a small high-speed wheel for internal grinding, a larger slower wheel for outside grinding, and a cup wheel for face grinding. All operations, except loading and unloading, were automatic, which resulted in a very substantial speed-up of production.[5]

Automatic sizing of the work had also appeared by 1905 as a means of increasing production rates. A. B. Landis had developed such a mechanism,[6] partly electrical and partly mechanical. A click on the feed was actuated by a rod bearing on the work to stop the machine at a desired point. After a first rough grind, the diameter of the work was read on a micrometer. The micrometer control on the feed was then set for a final grind to the desired dimension. Landis also pro-

3. Patents No. 1,017,880 and 1,017,881 both of Feb. 20, 1912.
4. A later monograph on the History of the Automatic Screw Machine will describe the influence of this very important machine tool in increasing production rates.
5. Colvin and Stanley, p. 28, and *Abrasive Industry,* 1927, p. 87.
6. A. B. Landis patent No. 781,434 of Jan. 31, 1905. See also patent of A. P. Steiner No. 1,677,307 of July 17, 1928.

FIG. 62 BRYANT CHUCKING GRINDER, 1908
(*Bryant Chucking Grinder Company*)

FIG. 63 PRATT & WHITNEY AUTOMATIC SIZING GRINDER, 1908
(*Colvin and Stanley*)

FIG. 64 HEALD INTERNAL GRINDER WITH AUTOMATIC SIZE CONTROL, 1927
(*Heald Company*)

vided compensation for the wear of the wheel, but by a hand adjustment.

In 1908 Pratt & Whitney built a grinding machine embodying this feature for cylindrical and taper grinding (Fig. 63). The sizing device and automatic control of the feed enabled fully automatic production of parts finished to uniform and exact size, regardless of wear of the wheel. The operator merely removed the finished part, put in another blank, and started the feed mechanism. This machine also provided for course feed of the wheel until the work was nearly to size, and then fine feed was automatically thrown in for the finishing operation. Cuts of from .000125 to .003 of an inch could be obtained through an electrically operated sizing device.[7]

The introduction of hydraulic feeds into the grinding machine had been made in 1902 by Brown & Sharpe. With hydraulic drive any machine tool is much more easily made automatic. In 1922 The Heald Machine Company brought out their internal grinding machine with automatic hydraulic drive. This type of machine speeded up the operation of inter-

7. Colvin and Stanley, *American Machinist Grinding Book,* New York, 1908, p. 23.

nal grinding, but the size of the hole was still checked manually by testing with a plug gage. On close tolerance work this might require several measurements on each piece, for which the grinding spindle would have to be run back clear of the work to enable the plug gage to enter. All this took time and naturally restricted output.

In the same year the Heald Company undertook the development of an internal grinding machine with automatic size control, to meet the rapidly increasing demands for higher production, greater accuracy, and lower labor costs in the automotive and other metal working industries. The result was, in 1927, their internal grinder with automatic size control as shown in Figure 64. In this fully automatic machine, provision was made for roughing and finishing traverse and grinding feed rates, automatic truing of the grinding wheel before the finishing operation, compensation for wheel wear, and automatic stop and start of the chuck and the coolant.[8]

In the 1920's the Landis Tool Company developed an important series of automatic grinding machines aimed especially at the needs of the automobile industry, thereby continuing the advances which had begun in the early work of A. B. Landis. By 1920 P. Stoner had designed a special crankpin grinder for six-throw cranks, intended to increase production rates.[9] In this machine (Fig. 65) two revolving chucks held the work in exact alignment, and power was applied to revolve the crankshaft from both ends. The offset weights of the crankshaft were counterbalanced at each step. The unground crankpins could be brought into the exact position for grinding without removing the crankshaft from the machine.

A design to make this type of machine fully automatic was patented[10] by 1935. It would grind the crankpins to uni-

8. The Heald Machine Company, *Yesterday, Today, Tomorrow,* Worcester, Mass., 1951, p. 12, and *Abrasive Industry,* Aug. 1925, p. 9 and Oct. 1925, p. 320. See also the patent of A. P. Steiner (of Landis) No. 1,668,213 of Mar. 13, 1928.
9. Patents No. 1,453,572 of May 1, 1923 (filed Nov. 18, 1920) and No. 1,658,539 of Feb. 8, 1928 (filed Jan. 13, 1922).
10. H. F. Klingele patent No. 2,014,768 of Sept. 17, 1935.

FIG. 65 LANDIS AUTOMATIC CRANKPIN GRINDER, 1923
(*Landis Tool Company*)

form diameter, and within given tolerances without handling the crankshaft, once it was put in the machine. The crankpins would all be ground completely automatically, and then the machine would stop. This machine could also be adjusted to take crankshafts of varying lengths. However, it was not until very recently that a machine to accomplish this goal has actually become practical.

The first automatic camshaft grinding machine[11] was brought out by Landis in 1928 (Fig. 66). They also developed an automatic internal grinder in the same year[12]. Soon after, Landis put on the market automatic grinding machines for the automotive industry, using hydraulic controls and hydraulic clamping and release of the work[13]. Many of these machines embodied A. P. Steiner's automatic sizing device[14] which controlled both rough and finish cuts.

Many of the specialized grinding machines we have already described, such as crankshaft and camshaft grinders

11. R. K. Rowell patent No. 1,675,466 of July 3, 1928.
12. P. Stoner patent No. 1,663,148 of Mar. 20, 1928.
13. A. P. Steiner patent No. 1,744,587 of Jan. 21, 1930.
14. Patent No. 1,677,307 of July 17, 1928.

FIG. 66 THE FIRST AUTOMATIC CAMSHAFT GRINDER, 1928
(*Landis Tool Company*)

for automobile work, were made multiple-wheel and fully automatic in order to get higher production rates. Some idea of the kind of grinding machines which have resulted today from the union of specialization with automatic action can be gained from the machine shown in Figure 67, which finishes the leading and trailing edges of jet blades, one every fifteen seconds.

Surface and Disk Grinding Machines for Production

Not only the familiar cylindrical and internal grinding machines were made fully automatic to bring about high-speed production grinding. Two other types of grinding machines, which we have seen well developed in the 1880's, had lain almost dormant for a generation: the disk and the surface grinders. These grinders had remained toolroom machines because of lack of demand for precision or hardened flat surfaces on a production scale. There were also technical reasons. Before the days of the planer and shaper, flat sur-

FIG. 67 GRINDING MACHINE FOR FINISHING JET BLADES (*Lewis*)

faces had been carefully chipped with a hand chisel and then filed. A hundred years ago the test of a good machinist was his ability to perform this operation with skill and even a kind of flourish[15]. Flat surfaces which required a close fit, such as the slide valves and crosshead slides of steam engines, were filed and then patiently hand-scraped to fit. The planer had replaced the chisel for this operation, but finishing by use of a file was still common in the 1880's. But as soon as a hardened or a precision flat surface was required, the file and scraper could no longer stand up.

Another problem which had limited the use of the planer and shaper and even the milling machine for production work was how to hold the work on the reciprocating table. The lathe and its related tools held the work in a chuck or between centers. But these other machines usually required some kind of fixture to be made by the toolmaker at

15. So long did this tradition persist that when the author first entered the machine-tool laboratory as an undergraduate at M.I.T. in 1926 the first project assigned was to true up and square up a block of cast iron with only a mill bastard file, a file card, a surface plate, and a try square. Our second project was to hand scrape a D-slide valve. The results were pitiful, but they did inculcate a life-long respect for the hand skill of the old-time machinist.

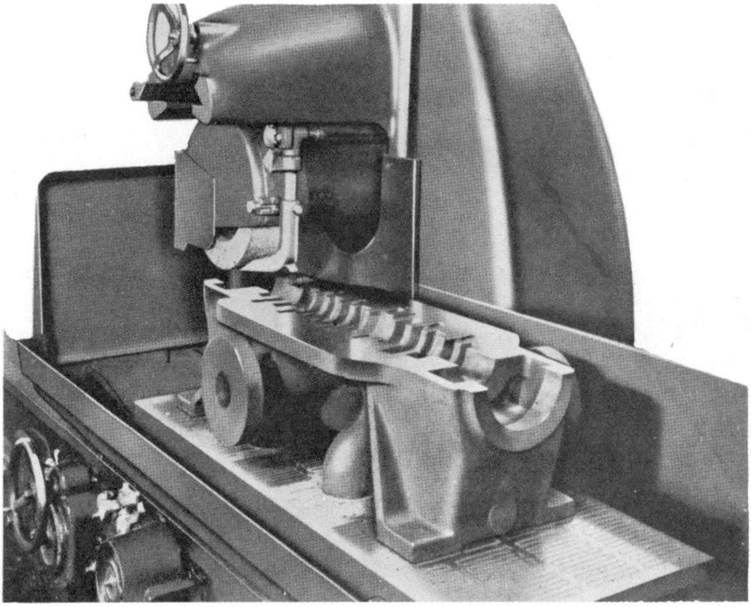

FIG. 68 MODERN SURFACE GRINDER (A 40-hour planing and scraping job was done in 3½ hours as shown) (*Lewis*)

considerable cost of time and money. With the invention of the magnetic chuck by O. S. Walker in 1896 a simple means of holding the work on the table was available, not for the heavy stresses put on the work by the planer, shaper, or miller, but more than adequate for the light touch of the grinding wheel. As we shall see in a moment, the twin-disk grinder and the various types of centerless grinding machines were to remove all need for direct support of the work and thereby allow automatic hopper loading to replace even that function of the workman. Surface grinding was even to affect the design of machine parts, for while planer and shaper tools work best on broad uninterrupted surfaces, since they tend to gouge in a bit as they strike a new edge, the grinding machine could save metal, time, and expense by working on a skeletonized surface.

Five basic types of production surface grinders were developed in the first generation of the 20th century. Four of these types had appeared much earlier and been developed

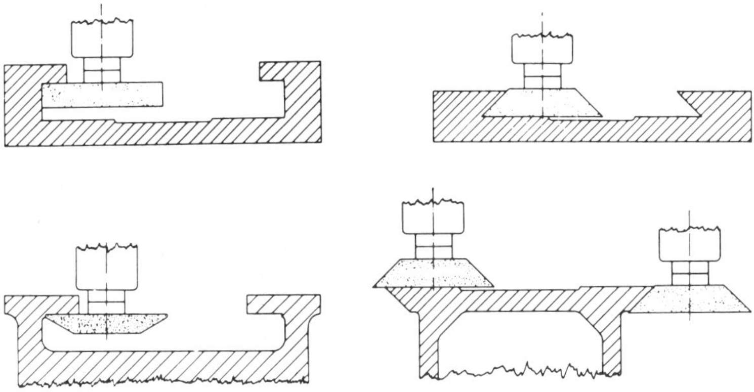

FIG. 69 GRINDING WAYS AND SLOTS OF MACHINE TOOLS ON A VERTICAL SPINDLE SURFACE GRINDER (*Lewis*)

into light toolroom surface grinders by 1887. As classified by Lewis, Type I has the axis of the grinding wheel horizontal and grinds on the periphery of the wheel moved under it on a reciprocating table (Figs. 24, 33a, 33b, 33c and Fig. 68). Type II has a similar mounting of the wheel, but has a rotary table for carrying the work (Fig. 57). The next three types all use grinding wheels of the cup, cylinder, or segmental forms. Type III has such a wheel mounted on a vertical spindle and grinds work on a reciprocating table. Figure 51 shows one of the planetary type. See also Fig. 69. Type IV is the face grinder, with a similar wheel mounted horizontally, but grinding vertical surfaces of the work carried in a reciprocating table. (Figs. 8, 22 and 33d). Type V has a similar wheel mounted on a vertical spindle, and its work is on a rotating table (Fig. 23). Each of these types has its special merits in speed, accuracy, finish, or convenience.

By means of types I and III it has been possible to grind with precision the ways, gibs, tables, and slots of machine tools, including grinding machines, more quickly and cheaply than before, and thereby avoid much of the expensive hand scraping previously used. However, scraping is still widely used for final finish of the ways of machine tools, because it permits a check and control on the distribution and extent of points of contact. It also produces minute oil pockets which help to lubricate the sliding surfaces and act as receptacles for small particles of gritty material that might otherwise damage the bearing surfaces. Type V surface grinding machines are especially suited to the production of extremely high precision in flatness and very fine surface finishes, as in gage blocks.

The surface grinder has also been made for high-speed production, some using multiple wheels and some in fully automatic models.

The disk grinder has developed into several types since it made the transition from the use of emery cloth to thick emery cloth, with grains "tilted" to improve their cutting. Finally about 1915, there was a transition to bonded abrasive disks attached by various means to a face plate, which made it virtually a face surface grinder classified as Type IV above.

The only type of disk grinder of interest to us here, since it was the only one used for precision production work, proved to be very important for high production rates. This is the double-spindle twin-disk grinding machine. This grinder has two abrasive disks with their faces opposite each other and the axes of their separate spindles coinciding. These two disks rotate in opposite directions, and their grinding faces are separated by a distance determined by the thickness desired in the piece to be ground. The work is fed between the two disks and the grinding of its opposite faces guided by a number of different techniques, some completely automatic. It should be noted that in this machine, as in the centerless grinders, one reason for their high production rates is that the work is not *held* in the machine while being ground, but is simply fed in from a magazine or hopper and then *guided* past the abrasive surfaces.

Piston rings, clutch faces, bearing races, brake drums, ends of helical springs, and a host of parts requiring precise thickness between two exactly parallel faces of fine surface finish, can be ground at high rates on the twin-disk grinding machine. Wrench forgings have as much as .030 of an inch of stock removed at a rate of 2400 units per hour, and with tolerances of .001 of an inch for uniform thickness and parallelism of the sides. Hardened piston rings have only .001 of an inch removed at a rate of 10,000 units per hour, with a tolerance of .0002 of an inch for size and .0001 of an inch for parallelism, and with an extremely high quality.

Clearly the surface grinder was doing for flat surfaces in production what cylindrical and internal grinders had done for cylindrical surfaces, and doing it with the same low cost and high rate of production.

Form and Thread Grinding — Crush Forming

Because grinding could deal so easily with hardened steel and because of its advantages for precision high-speed production, it was quickly adapted to work which had previously been done on other more conventional machine tools. Just as grinding had taken over many of the operations of the lathe, the boring machine, the planer, and the shaper, it also proved to be the best way to do some of the work previously machined on the gear cutter, the milling machine, and the screw machine. In addition to advantages we have already noted for the grinding process, the grinding wheel itself could be relatively easily formed into profiles in its cutting surface by the use of the diamond point. While at first these forms were difficult to maintain under production conditions, better methods of manufacture soon gave wheels able to retain the form trued in their cutting surfaces. Various means of renewing that form were quickly and easily developed.

The gear-cutting machine had by the end of the 18th century required the use of formed cutters, both single point and multiple, in order to form the epicycloidal or involute shape of gear teeth with speed and precision. The difficulty of maintaining the necessary profile on these cutters had led to Joseph R. Brown's patented milling-type gear-tooth cutter for easy sharpening. The needs of the teeth of bevel gears and convenience in sharpening the cutting tools were met by gear-cutting machines of the generating type. Very precise gears are still produced by gear cutters of tool steel or cemented carbides. But the needs of the automotive industry for hardened steel gears of strength, precision, and quiet operation at widely varying speeds and loads, soon required gears to be ground. Some machines were developed to do this by generating methods, but the grinding wheel with the necessary profile to form the teeth is still in common use. In either method it is necessary to have frequent and precise forming of the grinding surfaces of the wheel by a diamond-point tool guided to the exact form and mounted for easy manual or even automatic operation[16].

The milling machine of over 100 years ago was already doing form cutting, and special cutters for this work were designed in a host of shapes for general and special work. These cutters were, however, very expensive to make and difficult to maintain in correct profile in the sharpening process. Hardened steel could not, of course, be cut on the milling machine[17]. We have seen that as early as 1903 simple radius profiles were being formed on grinding wheels. Properly guided, the diamond point could form the desired profile in a grinding wheel and renew it quickly, easily, and cheaply as often as necessary. Form grinding of the races for ball bearings was in use in production by 1899 and 1906 in Pratt & Whitney specialized ball-bearing-race grinding machines, but the form was obtained by control of the motion of the grinding wheel by cams or a templet, rather than by forming a profile on the cutting surface of the wheel. The 1834 grinding machine by Wheaton had provided for cylindrical form grinding, but not by forming the cutting surface of the wheel. It was a simple matter to adapt these elements to production form grinding after 1900.

The enormous demands for screws and bolts of all kinds in the machinery of the 19th century had brought about the invention of specialized lathes intended to produce screws in tremendous volume—the automatic screw machine. Up to the 1940's all production screws were manufactured on batteries of these machines. Only taps, gages, and a few special devices had screw threads ground out of hardened steel[18].

The demands of the aircraft industry in World War II

16. R. S. Woodbury, *History of the Gear-Cutting Machine,* The Technology Press, Cambridge, Mass., 1958, pp. 121-124. For the special use of gear grinding in the automotive industry, see Fred B. Jacobs, *Production Grinding,* Cleveland, Ohio, 1922.
17. See the author's *History of the Milling Machine,* The Technology Press, Cambridge, Mass., 1960.
18. For a Swiss precision machine for making hardened screw-thread gages, using a single wheel formed by a precision-controlled diamond point and checked by a microscope, see *Abrasive Industry,* Nov. 1925, p. 359.

brought screw-thread grinding into production manufacture. Screw threads were required in material too hard (or too soft) for other methods, and screw threads were needed of much higher precision than could be obtained on the screw machine or lathe. Threads were ground to limits of .0002 of an inch on the pitch diameter, .0002 of an inch per inch of lead, and flank angles held to within five minutes of arc. In addition, thread grinding proved to be more economical.

Special thread-grinding machines have been designed to grind internal or external threads, or even both. And many automatic screw machines have been converted to thread grinders. The thread grinders are largely automatic for high production rates, and the operator merely loads and unloads the work. Thread grinders of the centerless type have been developed, which require only filling the loading hopper and emptying the finished work from another hopper[19].

Of course, all these machines must have precise and automatic truing of the wheel. A number of methods of doing this for thread grinding are in use for the fine-grained wheels necessary to retain precision forming. In one, a "single-rib wheel" has a single-face contour trued to the form of the desired thread, and the wheel is traversed as though it were a metallic cutting tool. There are several types of "multi-rib wheels"; some have a face as wide as the required length of thread and are form trued for making a single plunge cut with a single revolution of the work; others are formed so that successive ribs take roughing, intermediate, and finishing cuts in a traversing operation. The multi-rib wheels are usually resinoid to give the best surface finish, but vitrified wheels are used for highest precision. Oil is the usual coolant; because wheel speeds are much higher than in ordinary grinding, very careful wheel balancing is required. For the high accuracy required, the wheel is usually adjusted for the helix angle of the thread being ground. Of course, these machines are fully automatic, since they are designed for precision high-speed production (Fig. 70).

Although the diamond point still remains the standard

19. *Grits and Grinds,* Feb. 1958, p. 3.

FIG. 70 MODERN AUTOMATIC THREAD GRINDER (*Lewis*)

FIG. 71 CRUSH FORMING A GRINDING WHEEL (*Lewis*)

method of producing the desired profiles for form grinding, in World War II crush forming came to the fore to meet the demands of airplane and munitions manufacturers for fast and accurate forming of screw thread and other profiles on grinding wheels. This technique consists of using a hardened steel roll on which the desired profile has been reproduced. Rigidly mounted in a substantial housing, this roll is brought up to the grinding wheel. Both are then rotated under considerable pressure until the form on the roll is impressed into the grinding wheel (Fig. 71). Recently, cemented carbide rolls have been used for crush forming. The technique can be used only on vitrified wheels. The advantages and limitations of crush forming are too technical to be discussed here, but for many purposes crush forming is superior to diamond-point forming technically, and in all cases where it can be used it provides initial forming of the wheel profile in a few minutes and renewing of a worn formed wheel in a few seconds.

Centerless Grinding — External, Internal, and Thread

Centerless grinding has tremendous appeal for high production rates at low cost. It results in fully automatic machines, multi-machine group operation, and easily trained operators, a combination to delight the heart of a shop superintendent, even if he does have troubles with his highly skilled "set up" men.

The basic principle of centerless grinding is that the work is not directly supported nor directly rotated. Instead, it is rotated by the combined action of the regulating wheel and the grinding wheel and guided past the abrasive surfaces by a number of types of devices. Centerless grinding may be as old as Wilkinson's spindle grinder of about 1820. Schleicher's self-feeding needle grinding machine of 1853 (Fig. 25), is clearly a single-wheel centerless grinding machine, and Poole's roll grinder of 1870 is a kind of centerless grinding in reverse. There are patents on centerless grinding from 1874. The Detroit Machine Company advertised in 1912 a single-wheel type in which the work was guided by a feed block and ground by a wheel 16 inches in diameter and

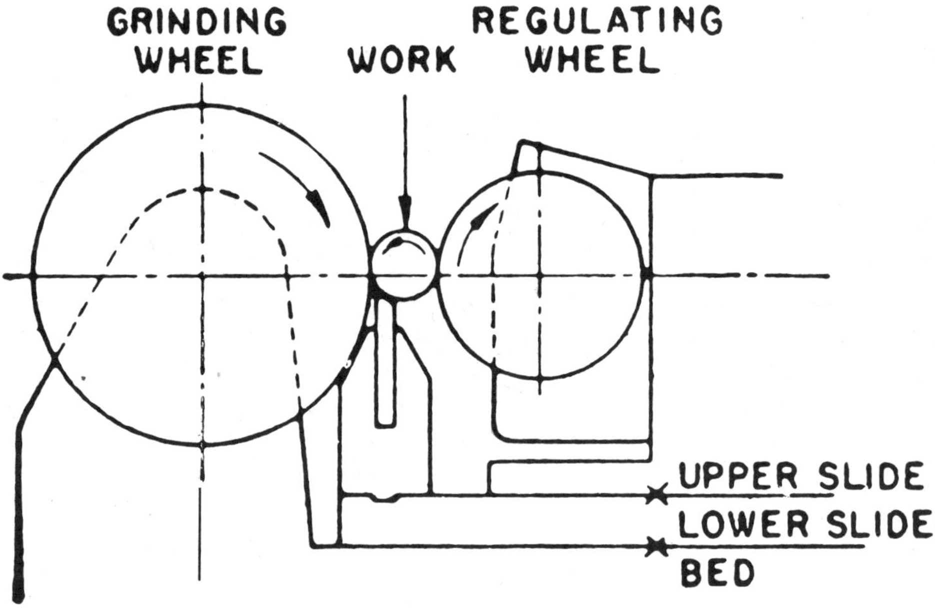

FIG. 72A PRINCIPLE OF CENTERLESS GRINDING (*Cincinnati*)

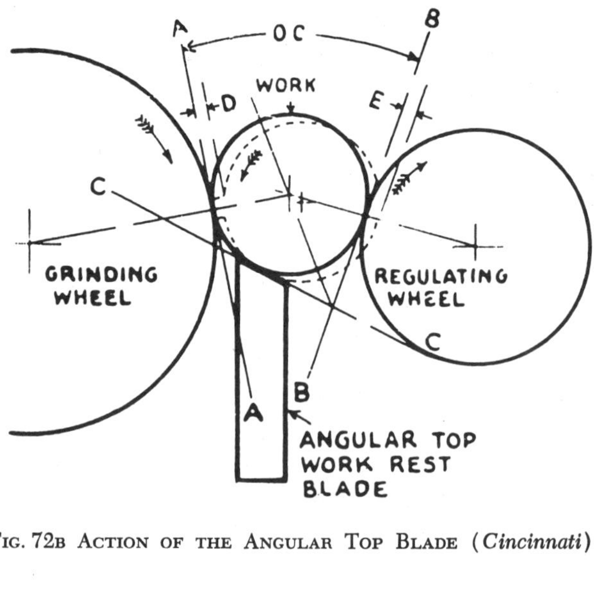

FIG. 72B ACTION OF THE ANGULAR TOP BLADE (*Cincinnati*)

4 inches wide. The wheel was trued slightly conical to pull the work through[20]. However, up to the time of World War I, centerless grinding had been limited to finishing shafting and the improvement of cold drawn surfaces.

The critical invention for centerless grinding was the regulating wheel and work blade by L. R. Heim in 1915[21]. The operation of this device is shown in Figure 72a. The work lies between two wheels, the grinding wheel, usually vitrified, rotating about 6000 feet per minute in the direction shown by the arrow, and the smaller regulating wheel, usually rubber bonded, and rotating much more slowly, in the opposite direction, at 36 to 900 feet per minute. The contact of the work with the grinding wheel rotates the work in the direction shown at a surface speed which would approach that of the grinding wheel were it not braked by its contact with the regulating wheel. There is therefore a difference in surface speed between the grinding wheel and the rotating work; this does the grinding.

It is, however, important that the work be supported between the two wheels in the correct position for grinding and to give the correct pressure against the regulating wheel. These functions are performed by the blade, as shown in Figure 72b. As will be seen, the work is supported with its axis slightly above the centerline of the axes of the two wheels. Since the work is seldom supplied perfectly round, it is necessary to have this difference in order to assure perfect roundness of the part when ground.

Successful centerless grinding depends upon careful adjustment of all these factors to give the fine finish and high accuracy of which grinding is capable, of the order of .0002 of an inch.

There are several methods of feed of the work. "Through feed" was the earliest and was used for grinding cold-drawn steel bars, at first to remove the hard skin on the surface to permit inspection for defects in cold drawing, later

20. *Abrasive Industry,* Feb. 1925, p. 64. See patent of R. H. Grant No. 1,106,803 of Aug. 11, 1914.
21. See patents No. 1,210,936 and No. 1,210,937, both of Jan. 2, 1917. See also E. L. French and G. W. Stephenson, No. 1,111,254 of Sept. 22, 1914. Mention should also be made of the early work of F. C. Sanford.

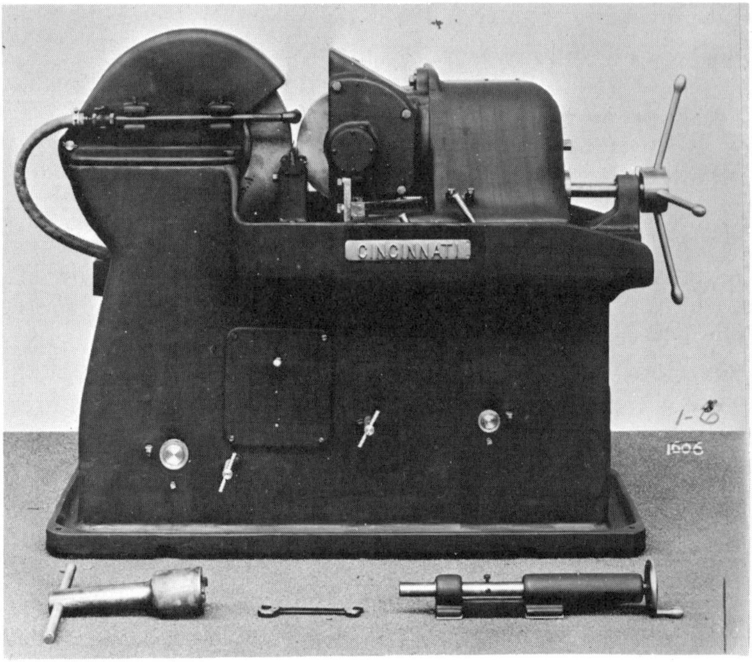

Fig. 73 Cincinnati Centerless Grinding Machine, 1922 (*Cincinnati*)

because of the resulting accuracy and lower costs. In through feeding the work is drawn through the machine by the action of the wheels produced by a slight skewing of the regulating wheel.

The advantages of high precision and high production rates possible by centerless grinding became evident in the early 1920's, largely through the initiative and vision of the Cincinnati Milling Machine Company. Their first production machine of 1922 is shown in Figure 73[22]. Although intended for general use, the immediate applications of this machine to automobile manufacture were indicated, since it could grind shoulder work, such as push rods and valve tappets. The grinding wheel was 20 inches in diameter, the regulating wheel 12 inches, and both had a 3-inch face. Sixteen changes

22. *American Machinist,* Aug. 30, 1923, p. 339. See also their pamphlet Form 1065, *The Cincinnati Centerless Grinder,* 1923.

FIG. 74 PROFILE CENTERLESS GRINDING (*Cincinnati*)

of speed were provided for, from a motor of 10 to 15 horsepower. Special care was taken to eliminate vibration. Diamond truing devices were incorporated for both wheels. The work blades had stellite wearing surfaces. It is from this machine that precision high-grade centerless production grinding originates.

Many applications to other types of work required "in feed" techniques. These are plunge-cut operations and therefore need wide wheels, but they do permit profile grinding, as shown in Figure 74.

The demand for greater width wheels and more powerful machines, [23] for in-feed grinding of long parts led by 1932 to the development of centerless grinders of the type shown in Figure 75. These machines were originally used for grind-

23. See patents of M. Romaine Nos. 2,065,099 and 2,065,100 both of Dec. 22, 1936.

ing cast-iron crankshafts for automobiles and were later applied to through-feed grinding of automobile engine pistons. These machines illustrate the adaptation of centerless grinding to highly specialized machines for producing specific parts.

By 1925 centerless grinding had developed so rapidly that it was grinding automobile valve stems 6 inches by ⅜ of an inch at the rate of 350 per hour, piston pins at 225 to 325 per hour, fountain pen barrels (including their tapers) at 750 per hour, shackle bolts at 400 per hour, and king bolts at 325 per hour[24]. The regulating wheel required truing once or twice in an eight-hour shift, the grinding wheel more often, but it was the centerless grinding machine which taught manufacturers that the wheel itself is an expendable item and its cost minor in high production rates.

Grinding by the centerless method proceeded so rapidly that the principal problem soon became the loading and unloading of the part. Already the cost of skilled grinding machine operators had been eliminated and high rates of production attained, but designers were still not satisfied. The usual procedure for unloading had been to back off the blade and the regulating wheel and allow the finished part to fall through, but designers were impatient with this delay, even when done automatically and with mechanical loading of the next piece. In order to get rid of even this slight loss of grinding time the cammed regulating wheel was invented (Fig. 76).[25] As is evident from the drawing, a loading period (*1* and *2*) is followed by a roughing (*3*) and a finishing grind (*4*) and an automatic drop unloading

24. *Abrasive Industry*, Apr. 1925, p. 102. For other applications, such as to the sewing machine, and for a discussion of its technical merits and limitations and an excellent summary of centerless grinding practice by 1925, see W. J. Peets, in *Mechanical Engineering*, Sept. 1925, pp. 695-700. Centerless grinding has by no means been confined to small or symmetrical parts. By the use of outboard bearings, automobile axle housings are centerless ground, and by the use of suitable fixtures, the round shanks of aircraft propeller blades are ground by this same technique. It has even been applied to the production of anti-friction freight car axles (*Machinery*, Feb. 1954, p. 3).

25. This was a Cincinnati development. See patents S. A. Strickland, No. 1,851,265 of Mar. 29, 1932 and G. Binns, No. 2,025,714 of Dec. 31, 1935.

Fig. 75 Cincinnati Centerless Machine for Grinding Automobile Crankshafts and Engine Pistons, 1932 (*Cincinnati*)

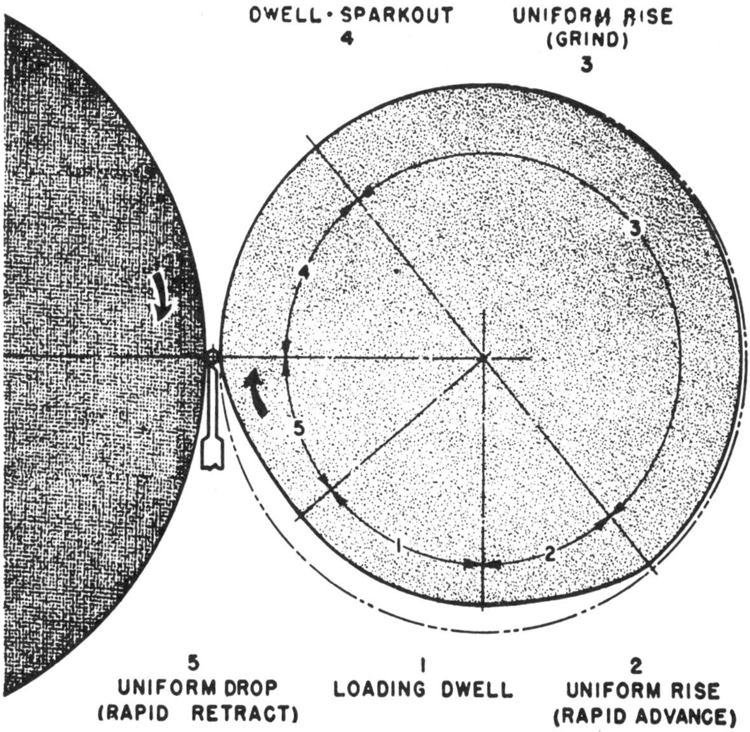

Fig. 76 Cammed Regulating Wheel. Centerless Grinding for Very High Uniformity and Rates of Production (*Cincinnati*)

FIG. 77 MODERN FULLY AUTOMATED CENTERLESS GRINDING MACHINE (*Cincinnati*)

(5). This device allows great simplification of the controls of the automatic centerless grinder, with consequent reduced initial cost of the machine. It also enables grinding with very high uniformity in the finished parts and the highest rates of production. Automation was also applied to centerless grinding. Figure 77 shows a modern fully automated grinding machine of this type, capable of grinding 150 automobile steering gear shafts per hour.

The internal grinder, even fully automatic and with automatic sizing, was a source of trouble in its chucking. The problem of holding the work to grind a hole in it was bad enough, and many shops spent more time working on the chucks of their internal grinders than they did using them to grind. But when machine designers wanted an internal surface ground precisely concentric with an external

Fig. 78 Heald Internal Centerless Grinder, 1933 (*Heald Company*)

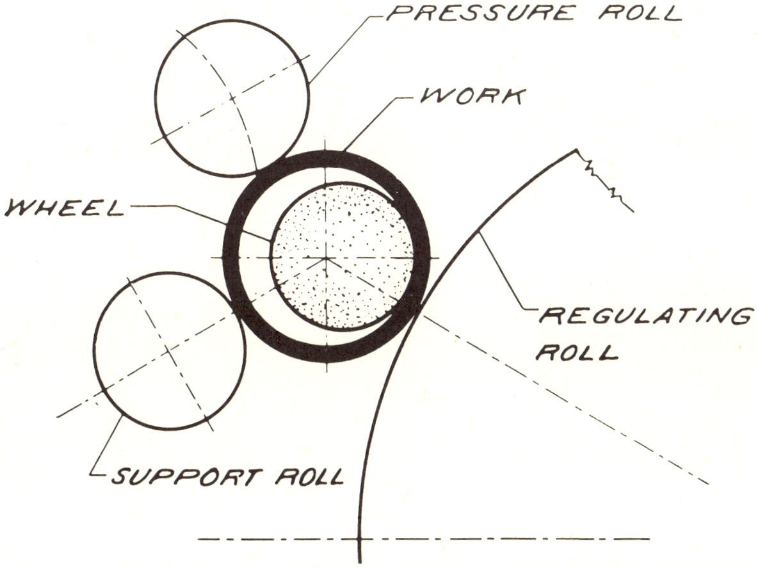

Fig. 79 Principle of Internal Centerless Grinding (*Cincinnati*)

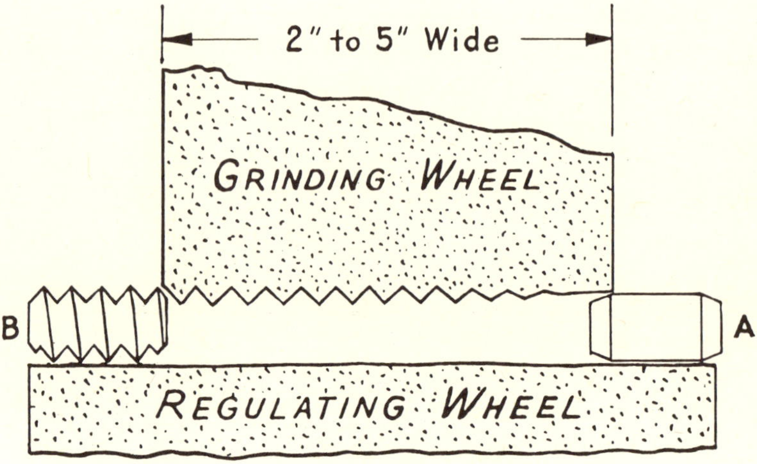

FIG. 80 CENTERLESS THREAD GRINDING (*Lewis*)

ground surface, shop superintendents threw up their hands. The internal centerless grinder was the answer, especially for the manufacture of anti-friction bearings, bushings, piston pins, and the like.

The Heald internal centerless grinder with automatic size control had been developed to meet this need by 1933 (Fig. 78). The basic principle on which it works is shown in Figure 79. The grinding wheel bears on the work exactly opposite the point where it is supported by the regulating wheel, and therefore there is no tendency to spring the work. Since the work is supported at three points on its already externally ground surface, the internal surface will be ground exactly concentric. Straight and taper holes, open or blind holes, interrupted holes, and pieces with external shoulders, can all be done by internal centerless grinding. These machines, too, are all fully automatic and have automatic sizing as well as automatic loading. A number of nice technical problems are involved in the very high speeds of their spindles and in the choice and supply of the grinding fluid.

Centerless grinding techniques have also been applied to thread grinding by the principles shown in Figure 80. In this development the Landis Machine Company[26] took the lead (Fig. 70). This method has been applied particularly to making hardened precision headless set screws. The screws are produced of precision quality with a single grinding pass at the rate of 2½ to 3 feet per minute, using crush forming of the wheel.[27]

High production rates of precise and hardened parts had been made possible at low cost by automatic machines, the application of surface grinding, form grinding, and by the many uses of the principle of centerless grinding. All this development put new demands on the grinding wheel and the technique of its use to meet the increasing calls for more and more precision and for better and better quality of surface finish.

26. Not to be confused with the Landis Tool Company of the same city. Their work stemmed from the U. S. patent of A. Scrivener of Birmingham, Eng., No. 2,427,024 of Sept. 9, 1947 (filed Sept. 29, 1944) and the later patents of C. W. Hopkins and others.
27. *Grits and Grinds,* Feb. 1958, p. 3.

MODERN GRINDING WHEELS AND SURFACE FINISH

1910 to 1950

The rapid increase in the use of grinding in industry raised many new problems and accentuated some old ones connected with the grinding wheel itself. New types of wheels were required with new characteristics, and there was an increasing demand for faster cutting and for abrasives to cut harder metals. Not only did demands for higher and higher precision continue, but the quality of the ground surface became an important question. New standards of surface quality resulted in increased use of honing, lapping, and even superfinish; and coolants and their application became far more important. Balance and vibration of the wheel were found to affect surface finish, especially at the high speeds required for internal grinding.

Improved Modern Grinding Wheels

By 1930 the search for new and better abrasives for use in production grinding wheels had begun. Diamond bort had had a limited and specialized use in grinding since the end of the 17th century, and the diamond point had been used for truing wheels, as we have seen. It was well known that the diamond was the hardest known substance, but it was expensive, even in industrial form. However, a new grinding problem arose: how to machine tungsten and other cemented carbides. The extreme hardness of these products of metallurgical research was of great advantage for certain uses, but the question remained of how to machine them. Baalis Sanford, of the Norton Company, took up the problem of making a diamond wheel to do the job. By October of 1930 he had made a tiny resinoid-bonded diamond wheel—only 3/8″ x 1/4″ x 1/4″. It worked very well in grinding Carbaloy, but the cost was prohibitive. Sanford continued his experiments, however, and in 1934 the Norton Company advertised a diamond wheel 7 inches by 1/2 inch.

Of course, the diamond bort was carried in only the outer 1/32 to 1/8 inch of the rim of the wheel. Within a few months the diamond wheel proved to be a commercial success as the only satisfactory means of machining the cemented carbides. By 1939 Van der Pyl, also of the Norton Company, had developed the metal-bonded diamond wheel for grinding hard non-metallic substances, such as glass, quartz, ceramics, and jewels. But until methods were later developed to recover the diamond bort, hitherto literally washed down the drain, diamond wheels were too expensive for most industrial grinding. Today in many grinding operations, the manufacturer cannot afford *not* to use diamond wheels. High initial cost, mitigated by recovery, in offset by savings in increased production.

As a result of the scientific studies on the grinding wheel which we shall examine in our next chapter, it became possible to produce abrasives having a more satisfactory structure through control of the grain size. On the one hand each grain had more cutting edges, and on the other the grains had a crystaline structure which would fracture and release new cutting edges just as the older edges began to get dull. This process was introduced in 1946 as the work of Raymond Ridgeway, of the Norton Company. By proper control of the fusing process in making aluminum oxide it was possible to form the crystals of a predetermined size, so that it was no longer necessary to crush the large crystals to get the grains desired for abrasive use. This process involved forming the crystals in the furnace in a sulphide matrix and then flushing out the matrix by the use of various chemical solutions, leaving the grain in finished form. When properly bonded into a wheel these gray-white abrasives proved to have cool, fast-cutting properties.

Advances were also made in the bond and the way in which it holds the abrasive grains. It is desirable for cool cutting to have the bond release the abrasive grain before it becomes dull. Better control of this bonding was obtained by improvement in the bond itself and by improved mixing of the abrasive grains with the bond in the new Hudson mixer, which wrapped each individual grain in its own coating of bond and did it with a substantial reduction in the time required for mixing.

Even more important was the recognition of the importance of the entire structure of the matrix in a grinding wheel—abrasive, bond, and air spaces. Wheels had been marked to indicate their kind and size of abrasive, their kind and strength of bond; and now it was found to be essential to indicate also the relation of their air spaces to abrasive and bond, or grain spacing. Techniques for controlled structure of the matrix were developed by Howe and Martin of the Norton Company in the 1930's. This process involved a very careful standardization of the weights of abrasive and bond for a given size of wheel, plus a proportional pressure put on the mixture to give the desired matrix. For surface grinding, an open structure and wide grain spacing provided adequate chip clearance, while wheels for form grinding needed a denser structure in order to give maximum strength to formed edges and corners, as in thread grinding.

Improvements in the manufacture of wheels resulted in more uniformity, as well as lowered costs. One such improvement was the replacement by 1919 of the old periodic kilns by the tunnel kiln. In this kiln the firing is a continuous process with the wheels passing through in a much shorter time and with much closer control of the firing. The manufacture of the abrasive itself was improved in the mid-1940's by the design of a production line in which most of the process was controlled merely by pushing buttons, from unloading the bauxite to loading the abrasive from the electric furnaces on to freight cars. About the same time, automatic and straight-line production methods were applied to making the wheels themselves.

All of these advances made available to the grinder better abrasive wheels and at lower costs.

Standardization and Safety of Grinding Wheels

By 1914 the number of types of grinding wheels being ordered from manufacturers had become so varied that there was a strong movement to have some kind of standardization, and a committee was appointed from within the industry to, "Cooperate with a committee of Machine Tool builders in order that:

1. The designers of grinding machines may take into consideration the standard grinding wheels already made and listed in regular catalogues.

2. To point out difficulties in making special shapes and the fact that all special shapes are not stocked.

3. Simplifying or eliminating, if possible, special shapes or sizes of wheels already catalogued.

4. Standardization of spindle diameter and methods of mounting as bearing on the use of countersinks and dovetails on special shapes or shapes other than plain wheels."[1]

What was desired was a code which would coordinate the design of grinding machines to utilize insofar as possible the standard sizes and shapes as then made. The report also pointed out the difficulty of manufacture and the increased costs of special shapes, and made a number of suggestions for standard safe mountings.

By 1920 the varieties of grinding wheels were estimated to be in excess of 715,000. In 1925 a conference was called by the Division of Simplified Practice of the National Bureau of Standards, which resulted in the first edition of the U. S. Department of Commerce, Simplified Practice recommendation R45, which managed to reduce the number of standard wheels to about 255,000. Of course, this figure included grinding wheels of all types and sizes in standard production, not just those used on grinding machines. In 1950, however, a grinding authority estimated that there were at least 16,000 types of wheels in regular use on grinding machines.

The standardization of the wheel itself was accompanied in 1918 by standard balancing practices, in 1920 by adoption of standard labeling symbols, and in 1922 standards for internal grinding wheels. Similar studies were continued in attempting to standardize what had been found to be good grinding practice.

1. *Abrasive Industry*, May 1925, p. 134.

Also in 1914 began the development of safety standards. By 1915 a Safety Code had been adopted. Over the years a number of tests were made to determine safe grinding practices. By 1943 a Standard Safety Code was well formulated, and a series of pamphlets was issued to instruct operators in safe means of grinding.

The details of both these standardization codes are of interest primarily to those in the industry. The mere fact that both developed rapidly after 1914 indicates that production grinding had achieved full stature in industry.

Grinding Fluids, Balance, and Vibration

From the beginning of grinding, one of its important advantages had been its special ability to produce a fine surface finish. As soon as a power-driven grinding wheel had appeared, it was very quickly noted that the coolant, used probably at first to keep the heat of grinding from drawing the temper of the work, also produced a better finish (Fig. 11). Even in Leonardo's day tallow was used on the grinding wheel. The fluids themselves have been water, water solutions of soda (to prevent rust), emulsions of oil in water, mineral oils, various paste compounds, and more recently certain synthetics.

The actions desired in these grinding fluids were several. First, it was important to keep the temperature of the work as uniform as possible and to prevent localized heating. This was especially important in grinding, since considerable heat is developed by the rapid removal of metal. Precision dimensions required that the work not expand appreciably while being ground, at least not in the direction of measurement. Grinding of thin tubes may distort them unless special care is taken with coolants.

Second, the grinding fluid acts as a lubricant to reduce friction between the metal and the abrasive as it cuts. This was less of a problem than with other types of metal-cutting tools, but there was no doubt that some lubrication improved grinding.

Third, grinding has a special problem. The chips of metal which it cuts are small; the grinding process releases used grains of the abrasive and other foreign matter. These must all be washed away so that they will not mar the surface of the work. Also, the metal chips must come free of the grinding wheel and not "load" its cutting surface.

And last, the grinding fluid must assist the wheel in producing the type of surface desired.

It will be seen that choice of a grinding fluid and proper methods of its application to the work at the point of grinding were of great importance, and became more so as heavy production grinding and high-speed precision grinding came into industry after 1900. Their importance was accentuated by demands for better surface quality in the 1940's.

It became important to filter the grinding fluids because the settling tanks of the grinding machines of 1915 were not adequate. The volume of fluid supplied became empirically stabilized at about two gallons per minute for each horsepower absorbed, and supplied from a tank holding 2½ times the flow per minute. However, for grinding to close tolerances, as in gages, the fluid is held at a standard temperature by automatic refrigeration.

Some grinding fluids have bad effects on the bonds of certain types of wheels, and only some are suited to the special needs of internal and centerless grinders. The value of fluids in nearly all types of grinding is unquestioned. The problem is just what fluid to use for a given job. Much remains to be done on this subject from the technical standpoint. From the production view there is no doubt that better surface finish and higher rates of production result from the use of the right fluid. And it is clear that fluids can cut costs by permitting the use of harder wheels without loss of cutting speed, thereby promoting longer wheel life, and at the same time making further savings by permitting heavier feeds.

We have already noted the high speeds required of the spindles of internal grinders and thread grinders. These introduce problems of balance and vibration which can quickly affect surface quality. These problems are accentuated in the internal grinder, which because it must be able to grind

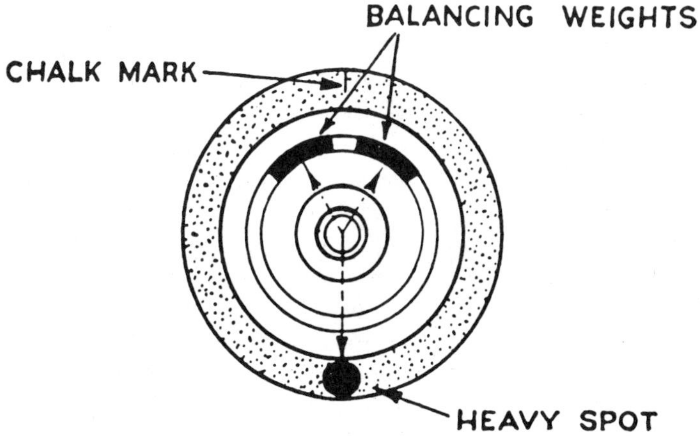

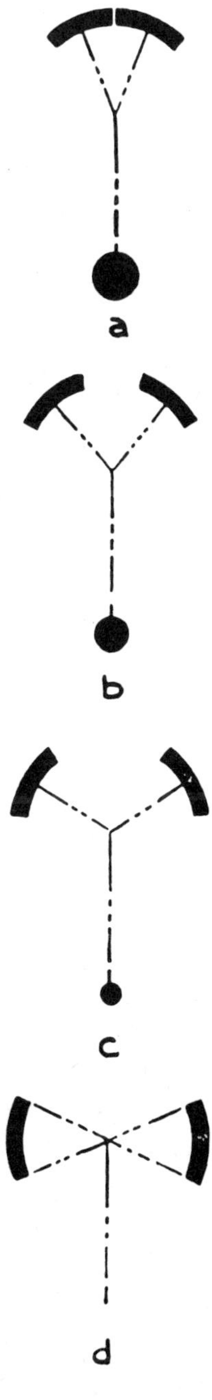

FIG. 81 PRINCIPLE OF ADJUSTING WHEEL BALANCE WEIGHTS (*Cincinnati*)

blind holes as well as open, must support a very high speed grinding wheel out on the end of a long spindle. The problem of adequate support for the spindle of the internal grinder had been met by Charles H. Norton in 1890 with his telescopic spindle. Further development was needed later, as might be expected.[2] For cylindrical grinders it was necessary to design special spindle bearings which would have zero vibration, yet be able to run cool at the speeds and under the pressures of heavy and high-speed production grinding.

By 1925 it was recognized that not all vibration had its sources in the machines itself. Belt drives could induce vibration, as could poorly fitted gearing.[3] In the 1930's better gearing and the application of direct drive by several motors eliminated these sources of vibration.

But there remained the wheel itself. Back in 1887 Charles H. Norton had found lack of balance of the wheel to be a cause of trouble in grinding. Lack of wheel balance not only leaves chatter marks on the work, but puts such a strain on the spindle bearings that matters soon get worse.

2. See, for example, the patent of H. T. Shearer No. 1,078,560 of Nov. 11, 1913.
3. *Abrasive Industry*, Feb. 1925, p. 44.

Although wheels were balanced at the factory very early, it proved necessary by 1925 not only to balance them again as mounted on their own sleeves for use in the grinding machine, but to rebalance frequently, especially after heavy grinding or substantial truing or forming. Special balancing flanges were developed to do this job. They had two adjustable weights, as shown in Figure 81. These weights were moved by trial and error into the position where they exactly offset the imbalance of the wheel, and were then held by set screws.

By 1950 a device had appeared to do this balancing of the wheel automatically while the wheel is still in place on the grinding machine.[4] The device is simple and compact and built right into the machine. Three balls, free to take positions in accordance with the forces acting upon them at operating speeds, settle into places which give perfect balance to the wheel. This condition is indicated to the operator by the needle of a vibration indicator, and he then locks the balls in place by throwing a lever. Rebalancing is then so quick and easy that the operator will do it frequently.

Surface Quality — Lapping, Honing, and Super-Finish

The desire to obtain high-quality surface finish on materials was not new in the 20th century. High surface polishes had been attained on glass for lenses and mirrors in the 16th and 17th centuries, as we have seen. We already noted Lassell's machine of 1851 for polishing mirrors (Fig. 23). The polishing of metal for armor was even older (Figs. 7, 11, and 12). But these operations were strictly polishing, the attainment of a high surface finish, but without concern that the surface itself was of precision dimensions, at least in metals. Nonetheless, polished surfaces for the sake of appearance were far more common in the machinery of the 19th century than they are today.

4. See A. B. Landis patent No. 1,091,851 of Mar. 31, 1914. Also see Cincinnati patents of H. Ernst, No. 2,142,021 of Dec. 27, 1938, J. C. Campbell, No. 2,164,900 of July 4, 1939, and A. H. Dall, No. 2,507,558 of May 16, 1950.

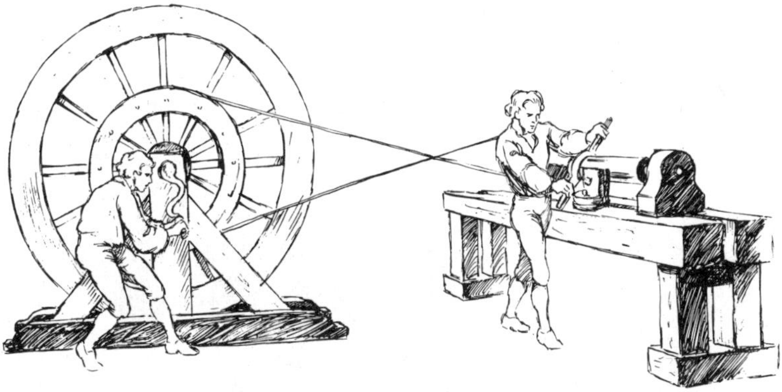

FIG. 82 LAPPING IN THE 18TH CENTURY (*Norton Company*)

Lapping is the process of using a free-floating grinding head to a get higher accuracy and a finer finish than are possible on cylindrical or flat surfaces by use of the grinding wheel. Honing is the same process applied to internal surfaces, although today generally done by a somewhat different technique. We have seen both processes before. In Figure 9 Leonardo da Vinci clearly used honing by 1500, and there is evidence that he used lapping as well. We find descriptions of lapping operations in the French encyclopedias of the late 18th century (Fig. 82). Honing was also in use to get good fits of metal shafts in solid bearings. Lapping was also used on the piston shafts of early double-acting steam engines in order to get the smooth surface and uniform diameter which would permit a tight fit of their stuffing boxes and so prevent the escape of steam. Peter Cooper of New York had a vertical lapping machine in 1835, and a power lapping machine was offered in Brown & Sharpe's catalogue as early as 1893.

Both lapping and honing are designed primarily to provide a better surface finish. They remove the surface waves, tool marks, grinding fuzz, slight distortions and like minor defects left by a preceding operation. Both the lapping and the honing tools are therefore not firmly held relative to the work, but float on it.

In lapping, the tool is usually of cast iron or other soft

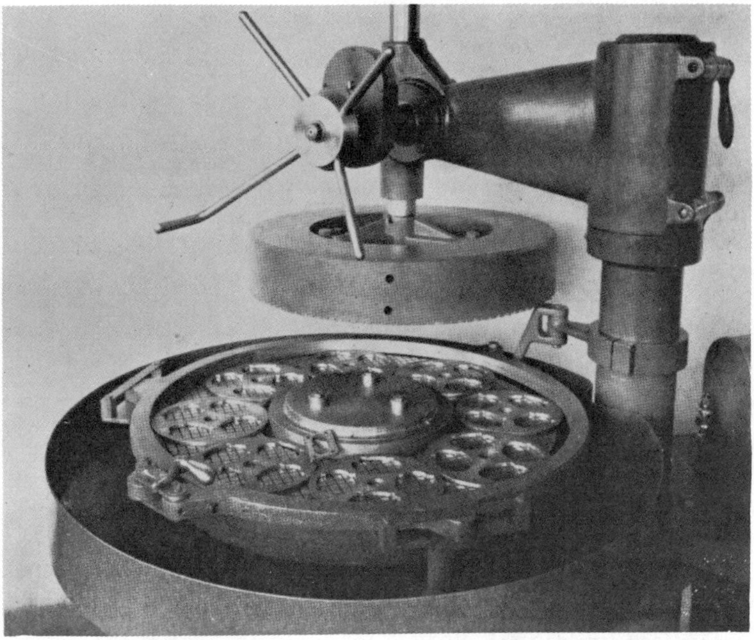

FIG. 83 MODERN VERTICAL LAPPING MACHINE (*Lewis*)

metal, the principle being that the lap, softer than the work, will charge itself with the abrasive and act as a grinding tool. Precision lapping requires that the abrasive grains be very uniform in size, otherwise scratches will result from the large grains and the small grains will do no work. A grinding fluid is necessary to lubricate the work, to suspend the abrasive, and to keep the grains separated. Most lapping machines for cylindrical and surface work are vertical (Fig. 83), with a cast iron bottom lap driven at about 300 feet per minute at the periphery, and a non-rotating upper lap which floats on the work, which in turn rests on the lower plate. The heavy upper lap supplies the grinding pressure on the loose abrasive grains generally used in lapping. The illustration shows the lapping of flat surfaces, in which it is necessary to make some provision to ensure uniformity of lapping by eccentric rigs so that the work moves from place to place on the bottom lap. For cylindrical work rolling of the pieces is required and is usually achieved by guiding the work in a special plate or cage such that the axes of the cylinders do not pass through the center of the lap.

Most vertical lapping machines can do either cylindrical or surface lapping; their product has extreme parallelism and straightness and an exceptionally fine finish. Lapping also permits very fine size control—on regular work to .000025 of an inch, and on gage work to .00001 of an inch.

Lapping machines have also been designed to use bonded abrasive disks, both upper and lower; and centerless lapping machines are in use. The lapping machine provides the very high precision, the flatness or straightness, as well as the surface finish required in certain modern machinery.

For lapping the crankshafts and camshafts of automobile engines special machines have been devised, using strips of abrasive tape. The first practical lapping machine for this purpose was designed by Herbert S. Indge to lap the wrist pins of the Wills-Ste. Claire car. This machine was the first to apply to a commercial product the surface refinements which up to that time had been used only for precision gages. In 1927 Indge joined the Norton Company, which brought out a commercial machine in the following year. Since that time fine surface finish has been applied to many other parts of the automobile, such as the hydraulic transmission, and to such things as the compressor pumps of electric refrigerators, the most precise mechanism in common use today.

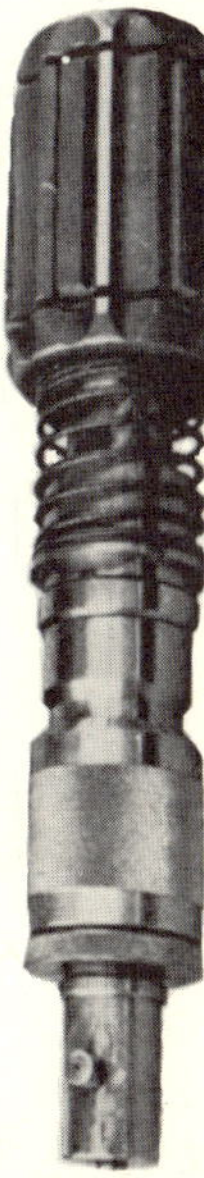

Fig. 84 Modern Honing Head (*Lewis*)

Honing is essentially lapping an internal surface, usually that of a cylinder. However, in honing we use abrasive sticks set in a honing head (Fig. 84). The sticks are forced out against the work by mechanical or hydraulic devices. This honing head is rotated in the cylinder and at the same time given a reciprocating motion, but it floats freely, guided only by the cylinder walls. Honing of cylinders came in originally as used in giving fine finish to the cylinders of French racing car engines. By the 1920's it was being used in the growing automobile industry.

Honing machines are usually vertical in order to minimize the effect of gravitation in tending to pull the hone to one side of the cylinder. The fact that interruptions in a surface being honed also produce irregularities in the honing action has affected the design of automobile cylinders to avoid this effect. Automatic size control is common, and honing produces an accuracy and finish of the same order as lapping.

The honing process has also been applied to the bore of big guns and to the cylinders of diesel engines. Without it the modern aircraft engine would not be possible.

Superfinishing is a process introduced by the automobile manufacturers about fifteen years ago. It was found that in cars shipped long distances by freight while still standing on their own wheels, the front wheel bearings were seriously damaged by the jiggling up and down of the balls on their races. Since these parts had already been given very close fits, this was at first difficult to understand. However, the cause was found to be a surface layer of fuzz, smear metal, or the like. When this was removed by especially fine lapping and honing machines designed to take parts already precisely finished to dimension and give them an extremely fine superfinish, the trouble disappeared.

The demands of the automotive industry for high-quality surface finish soon spread to other industries, and today drawings and blueprints have, in addition to dimensional symbols indicating high precision, new sets of symbols indicating the exact type and quality of surface finish which the machine designer requires.[5]

5. The story of precision means of measurement of surface finish in micro-inches will be told in a later monograph on the History of Shop Precision of Measurement.

THE SCIENCE OF GRINDING

1900 to 1950

In the first decade of the 20th century grinding took its place on the production line. In its earlier uses as a toolroom machine, or as employed in light industry, it had been possible to get adequate results from only the empirical knowledge accumulated by a few practical men who believed in the possibilities of grinding methods. But after the turn of the century the application of grinding to heavy work and to precision and high-speed mass production gave rise to problems—technical, production, and economic—which could not be met adequately by the practical experience and good judgment of men like Joseph R. Brown and Charles H. Norton, geniuses though they were. Since the middle of the 19th century all forms of machinery and many industrial processes had became less and less rule-of-thumb, and more and more engineering design. By 1900 this influence was being felt in the machine-tool industry, and the grinding machine was no exception.

With grinding entering so significantly into the mass production of the automobile, the stakes were too high to depend solely upon experience and good judgment in the design and use of machine tools for the precision and high-speed work required. We therefore soon find the tool engineer in the production shop. The problems of the more familiar metal-cutting machines were met for a while longer by the practical men,[1] but grinding soon required the special knowledge and skills of the trained engineer. And grinding had already become so important in certain industries that they could afford, in fact could not do without, his services.

1. But see the author's *History of the Milling Machine,* The Technology Press, Cambridge, Mass., 1960, for the experimental work on the milling machine done about 1908 by A. L. DeLeeuw at Cincinnati, John Parker at Brown & Sharpe, and P. V. Vernon in England.

Although the techniques of engineering science have been applied simultaneously to nearly all aspects of grinding since 1900, it will be convenient to examine them successively as focused on the grinding machine itself, the grinding wheel, and the abrasive cutting action.

Science and the Technique of Grinding

The application of engineering analysis to the grinding machine and its use was begun by James J. Guest of England, and the results were included in his book[2] of 1915. Guest had had a wealth of practical experience in all forms of grinding, as well as a very successful career as a designer of grinding machines of many different types. He was, moreover, able to turn his outstanding engineering mind to the problems of grinding.

He analyzed the problem of strength and surface speed of the wheel as affected by varying diameters. He discovered a formula relating the diameter of the wheel, the diameter and surface speed of the work, to the diametral reduction of the cut, to show that at one limit the wheel disintegrates too rapidly, causing waste and failing to size the work properly, but at the other limit the wheel glazes and fails to cut.

He calculated the amount of distortion which might result under various circumstances from dry grinding, and showed the other merits of grinding fluids. Guest was particularly concerned with the design of the bearings of the wheel spindle to give precision and freedom from vibration. This problem also involved the theories and practices of lubrication which had resulted a generation earlier from the researches into the nature of lubrication made by Osborne Reynolds and Beauchamp Tower, as extended in 1903 by Lasche to high-speed bearings.

Guest noted that at the high speeds required for internal grinding spindles the oil must be forced in. He used basic engineering principles for the special problems of these spindles, not only their lubrication, but also the attainment of

2. James J. Guest, *Grinding Machinery*, London, 1915.

the necessary rigidity and freedom from vibration for precision work and good surface finish. In particular, he analyzed in mathematical form the problems connected with the design of ball bearings for these spindles.

Even the belt drives of grinding machines came under his engineering analysis to good effect. Guest also designed a notable cross feed mechanism which gave less chance for error in bringing the wheel rapidly into action in repetitive work.

Guest was not merely a technical engineer; he was well aware of the importance of time and costs. He made a detailed analysis of these factors in grinding based upon data from Brown & Sharpe, from Landis, and from his own experience, and was able to prove what Charles H. Norton had advocated, that the cost of the wheel can largely be ignored, and that the cheaper element of power should be substituted for the more expensive one of labor, especially if softer wheels are used. Many production grinders used hard wheels in those days simply because they lasted longer.

Guest also established by mathematical analysis Norton's old idea of the value of using large and wide wheels, and he produced curves of the actual relation between work speeds and the cutting action of the wheel, which showed the importance of providing easy means of changing the work speed, especially on internal grinders.

Guest laid the foundations upon which later tool engineers produced sound design of grinding machines and effective use of them in practice.

One outstanding American who carried this work further was George I. Alden, of the Worcester Polytechnic Institute. In his study "Grain Depth of Cut,"[3] Alden proved mathematically the relationship between work speed, wheel speed, work diameter, rate of cross-feed, and the depth of cut to be expected from each grain, and therefore the principles on which to choose these elements for maximum production.

3. Presented at the Annual Meeting, Dec. 1914, American Society of Mechanical Engineers. Reprinted in Charles H. Norton, *Principles of Cylindrical Grinding,* Worcester, Mass., 1921.

In 1925 Alan A. Wood published the results of his studies and experiments on the surface grinding machine,[4] in which he showed the value for good chip clearance of the cup type of wheel. He also noted that the rapid cutting of this type of grinding machine arose on the one hand from the larger number of abrasive grains cutting at any one time, and on the other from the fact that the rigidity of these machines permitted the use of considerable pressure on the cutting edges. He also recognized that as a result these machines had extra coolant requirements to handle the greater heat generated. Wood went on to show how his experiments on the design of machine parts for the grinding operation indicated that recessed surfaces required far less power for grinding.

Science and the Grinding Wheel

We have already described some of the improvements in the grinding machine which resulted by the 1930's from a scientific attack on its problems. The Norton Company had established a small research department by the turn of the century, which grew in size and importance rapidly by the time of World War I.

With the discovery of the artificial abrasives and the fact that the characteristics of their crystals could be changed widely by various means, especially for aluminum oxide, the way was opened for scientific studies to improve not only the abrasive, but the bond and the whole matrix structure of the grinding wheel.

The control of the grain size and structure in the process of making aluminum oxide resulted from a number of scientific studies of just what happened in the furnace. It was noted that under certain conditions the crystals were quite large and therefore had to be crushed to give the size of grain desired for an abrasive wheel. The fragments which resulted had cutting edges and fracture strength quite different from whole crystals of the same size. A somewhat

4. *Abrasive Industry*, Dec. 1925, p. 376.

different type of aluminum oxide could be produced, also in crystals of large size. The abrasive grains from this material were, however, much weaker from perforations in the crystals caused by the presence of a vapor at the moment the crystal solidified from the molten state. This vapor was found to be caused by the presence of small quantities of alkali which were volatile at the furnace temperature. A weakening of the grain of the crystal could also be produced with predetermined exactness by introducing certain chemicals into the fused mass, or by a rapid cooling of the mass as taken from the furnace.

Certain forms of the grains used as abrasives were found to be made up of more than one crystal, and this crystal structure and grain make-up could be changed in toughness or nature of fracture by appropriate means.

Most of this information about grain and crystal structure was obtained by the use of the petrographic and metallographic microscopes and the X-ray spectrometer. Chemical analysis and the growing science of crystallography suggested ways in which these properties could be changed and controlled in manufacture.

Means were developed for determining accurately the hardness, toughness, and strength of abrasive grains. Hardness had previously been measured on the Mohs scale of scratch hardness from 1 to 10 (the diamond). Physicists were struck by the fact that as great a range of hardness existed between 1 and 9 as between 9 and 10, and that with the exception of the artificial silicon carbide and boron carbide, there were no abrasives to be found between 9 and 10. These facts suggested a better scale based upon a better test—the Knoop scale. This scale measures the micro indentation produced by a diamond point ground to a pyramidal form and used to indent the substance under varying loads. The Tukon tester developed by the Bureau of Standards was adapted by the research staff of the Norton Company for determination of the Knoop hardness of abrasive materials.

Science and Abrasive Metal Cutting

At the turn of the century Charles H. Norton, Brown & Sharpe, and others had studied the chips produced by the grinding wheel and recognized that properly used, the grinding wheel was in many respects like other metal cutting tools, except that its cutting edges were much harder and that there were thousands of these tiny cutting tools in use at a given time. In 1915 Guest had made some very shrewd observations of the probable nature of the cutting action of abrasive grains, not all of which have been borne out by later experimental work, but at least Guest recognized the importance of this question.

Only in the last twenty years have scientific studies of the exact nature of the action of the cutting tool in forming chips of metal been undertaken by means of specialized applications of metallurgy and the theory of the solid state of metals.[5] But a few things especially relevant to grinding have been discovered.

One rather puzzling set of observations has led to the recognition of a new element in the nature of the action of the abrasive grain in cutting metal. It was noted that although silicon carbide was clearly harder than aluminum oxide, the results in actual grinding sometimes seemed to belie this fact. Aluminum oxide is softer than silicon carbide, but it is a much better wearing tool on high and medium carbon steels. Yet silicon carbide has superior wearing qualities on the hard and brittle chilled cast iron, and also on copper, aluminum, soft bronze, and other non-ferrous soft metals. How to explain these facts? The theory advanced was first given a name—attrition. There appears to be in the action of an abrasive not only hardness and strength to be considered, but also a third factor, the ability of the abrasive to resist dulling on the work in question. This property has not been measured, but seems not to be related to hardness, at least as commonly measured. The theory advanced to explain attrition is that the abrasive material,

5. This development will be considered in a later monograph on the History of the Metal-Cutting Process and Its Tools.

in addition to cutting, also dissolves or diffuses into the material being ground, at the point of contact, because of the heat and pressure at that point. This theory has been given some support by the results found in grinding titanium. This substance, known in its molten state to react very vigorously with all known abrasives, uses up grinding wheels at a terrific rate.

The next few years will undoubtedly find science telling us far more about how to use the grinding machine to take the best advantage of its special characteristics, and may even point to uses we do not yet dream of.

CONCLUSION

The grinding machine began as a machine tool intended primarily for shaping and polishing substances too hard for other means. By 1865 it had grown into a toolroom and light manufacturing tool capable, in addition, of giving high precision. After the turn of the century it was adapted to heavy production, to high speed and automatic methods, and to machining parts of high surface finish and ultra precision.

The application of this machine in both standardized and specialized form to a number of industries has been examined, especially to the automobile industry, which was to become the largest single branch of our metal-working industrial economy. The grinding machine has made work which was previously impossible, easy and inexpensive. It has created the modern view of what constitutes high-class production machine work. Like the milling machine, it has taken over some work previously done on other machine tools, but it has far from supplanted them. In fact, the grinding machine has achieved much of its present importance by creating tooling operations and even industries which would not otherwise have been possible.

While much of our industrial economy is not based upon the principles of mass production, some very important segments of it are. In, for example, the automotive industry, interchangeable parts are of the very essence. The old-time machinist would be appalled to see the workmen in a modern assembly line putting together parts assembled from plants all over the country, and calmly assuming that of course they fit, and perfectly too. Underneath all this interchangeability lies precision. The grinding machine showed the way to achieve precision, in both hardened and unhardened parts, quickly and easily. This characteristic allowed the machine designer to do many things that he could not dream of even a generation ago, and in return he has found it valuable to keep the special requirements, as well as the advantages, of the grinding machine firmly in mind in making his design.

But all these technical advantages of grinding would have had only limited application and importance had not grinding been able to meet the competition of other machine tools on the basis of cost. Grinding showed conclusively that it could bring about increased productivity at reduced cost, as can be seen in the following table[1] showing a typical increase in productivity, in this case for grinding valve tappets for automobile engines. Since these tappets had to be made of hardened steel they could not be made at all on the older machine tools. This work was done to tolerances of .0005 of an inch round and straight in 1920, and to .00025 of an inch in all the later years.

Up to	*Operation*	*Pieces Per Hour*
1920	Center-type grinding	90
1923	Centerless in-feed grinding; hand operated, chucked, and ejected	150
1924	Centerless in-feed grinding; hand operated and chucked, lever ejected	240
1925	Centerless in-feed grinding; hand operated and chucked, automatic ejection	300
1927	Centerless in-feed grinding; automatic operation, chucking, and ejection	450
1929	Centerless in-feed grinding; hopper supply, automatic operation, chucking, and ejection (One man operating three machines)	1350

The grinding machine is the most significant addition to our production machine tools since Maudslay discovered the idea of the precision machine tool at the end of the 18th century. It has not only had an important influence on all production metal working industries, but it made technically possible[2] our great automotive and aircraft industries which not only changed the entire nature of our industrial economy, but brought profound changes in the way in which we all live.

1. Taken from S. Einstein's very important article, "Machine Tool Milestones—Past and Present," *Mechanical Engineering,* Nov. 1930, p. 959.
2. For a fascinating and scholarly account of the personal, organizational, and financial factors which created the automobile industry, see my colleague John B. Rae's recent book, *American Automobile Manufacturers, the First Forty Years,* N. Y., 1959.

BIBLIOGRAPHY

Books

APPLETON. *Cyclopedia of Applied Mechanics,* New York, 1882.

ACHESON, EDWARD G. *A Pathfinder,* New York, 1910.

BESSON, JACQUES. *Theatrum instrumentorum et machinarum,* Lugdini, 1578.

BÖCKLER, GEORG ANDREAS. *Theatrum machinarum novum,* Nürnberg, 1661.

BROWN & SHARPE CO. *Treatise on Grinding Machines,* Providence, R. I., 1891.

BYRNE, OLIVER. *Handbook for Artisans, Mechanics and Engineers,* Philadelphia, 1853.

CARDANO, GIROLAMO. *De rerum varietate,* Basileae, 1557.

CINCINNATI MILLING MACHINE CO. *The Cincinnati Centerless Grinder,* Cincinnati, Ohio, 1923.

COLLIE, MURIEL F. *The Saga of the Abrasive Industry,* Greendale, Mass., 1951.

COLVIN, FRED H. & STANLEY, F. A. *American Machinist Grinding Book,* New York, 1908.

DA VINCI, LEONARDO. *Codex Atlanticus*
Institut de France MSS.

DIDEROT, DENIS. *Encylopédie,* Paris, 1763.

FELDHAUS, FRANZ *Die Technik der Vorzeit, der Geschichtlichen Zeit und der Naturvölker,* Leipzig, 1914.

GUEST, JAMES J. *Grinding Machinery,* London, 1915.

HALSEY, F. A. *Methods of Machine Shop Work,* New York, 1914.

HEALD MACHINE CO. *Yesterday, Today & Tomorrow,* Worcester, Mass., 1951.

JACOBS, FRED B. *Production Grinding,* Cleveland, Ohio, 1922.

LEWIS, KENNETH B. *The Grinding Wheel,* Concord, N. H., 1951.

MARSH, E. A. *Evolution of Automatic Machinery,* Chicago, 1896.

NORTON, CHARLES H. *Principles of Cylindrical Grinding,* Worcester, Mass., 1921.

NORTON COMPANY. *Salute! Mr. Norton,* Worcester, Mass., 1941.

PRECHTL. *Technologische Encyklopedie,* Berlin, 1838.

RAE, JOHN B. *American Automobile Manufacturers, the First Forty Years,* New York, 1959.

ROE, JOSEPH W. *English and American Tool Builders,* New Haven, 1916.

ROSE, JOSHUA. *The Complete Practical Machinist,* Philadelphia, 1899.

SCHLESINGER, G. *Die Werkzeugmaschinen auf der Weltausstellung in Lüttich, 1905,* Springer, Berlin, 1906.

SCHROEDER, ALF. *Entwicklung der Schleiftechnik bis zur Mitte des 19. Jahrhunderts,* Hoya-Weser, 1931.

SMILES, SAMUEL. *Lives of the Engineers,* London, 1861.

STRADA, JACOB DE. *Künstliche Abriss allerhand Wasser, Wind, Ross und Handt-Mühlen beneben schönen und nutzlichen Pompen,* Frankfort, 1618.

TYMESON, MILDRED McC. *The Norton Story,* Worcester, Mass., 1953.

U. S. PATENT OFFICE. *Repertory of Patent Inventions,* Washington, D. C., 1832.

WITTMANN, KARL. *Die Entwicklung der Drehbank,* Berlin, 1941.

WOODBURY, R. S. *History of the Gear-Cutting Machine,* Technology Press, Cambridge, Mass., 1958.

ZONCA, VITTORIO. *Novo teatro di machine et edificii,* Padua, 1607.

Journals

Abrasive Industry

American Machinist

Civil Engineers and Architects Journal

Dingler's Polytechnische Journal

Grits and Grinds (Technical Publication of the Norton Company)

Farmers, Mechanics, Manufacturers and Sportsmans Magazine (New York)

Journal of the American Ceramic Society

Journal of the Franklin Institute

Journal of the Society of Arts (London)

Machinery (New York)

Mechanical Engineering (New York)

Scientific American

Tool Engineer

Transactions, American Society of Mechanical Engineers

Transactions, Institution of Mechanical Engineers

Transactions, Rhode Island Society for the Encouragement of Domestic Industry, (Providence, R. I.)

Transactions, Royal Scottish Society of Arts

Other Sources

Patent Files. U.S.A., Great Britain, France, Germany. (All references are to U.S. patents unless otherwise indicated.)

Industrial Files. As usual, the files of Brown & Sharpe have provided a wealth of material. A rich store of source material is also to be found in the files of the Norton Company. The files of The Cincinnati Milling Machine Co., Cincinnati, Ohio, also contain much material on the development of centerless grinding.

INDEX

WHEEL DIMENSIONS

WHEEL DIMENSIONS		PRECISION (*inches*)	CUTTING CAPACITY (*cu. in. / min.*)
	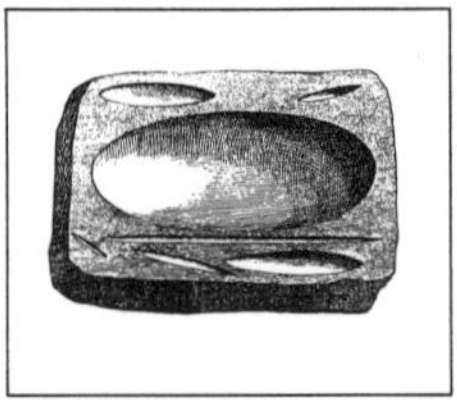 Prehistoric	 0.020	0.004
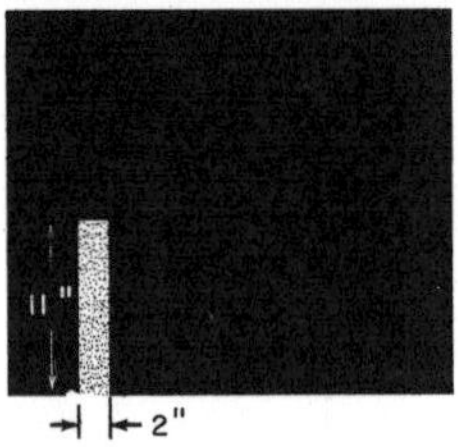	 About 1800	 0.020	0.020
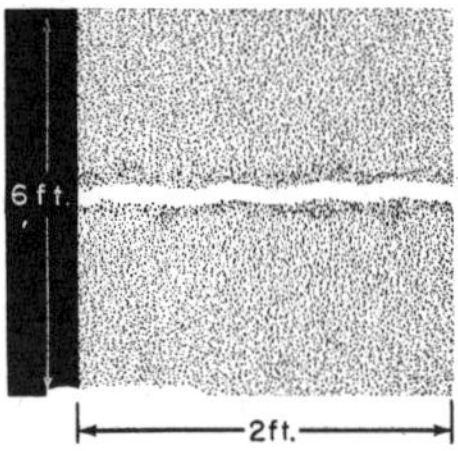	 About 1820	 0.100	0.100
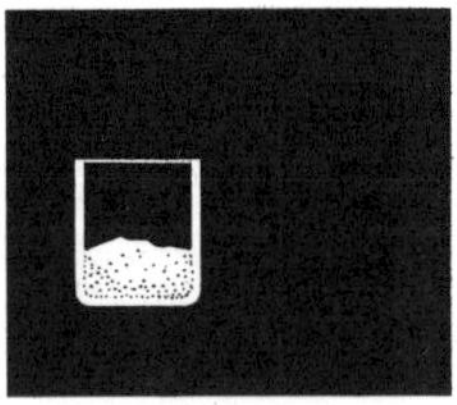	 About 1840	 0.002	0.001
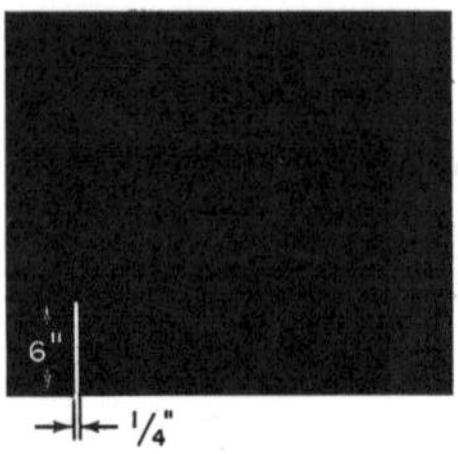	 About 1860	 0.005	0.020

WHEEL DIMENSIONS

PRECISION (*inches*) — **CUTTING CAPACITY** (*cu. in./min.*)

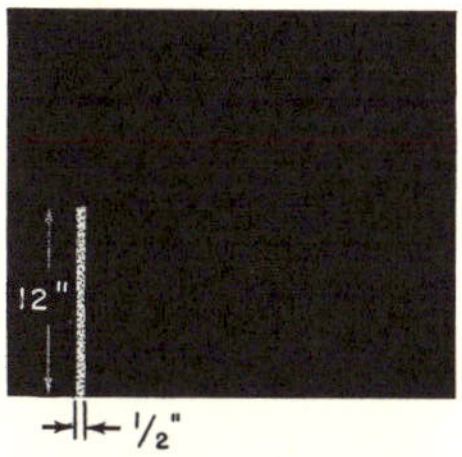

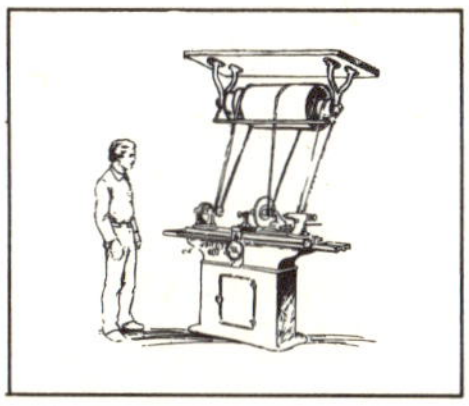
About 1880

0.002 — 0.060

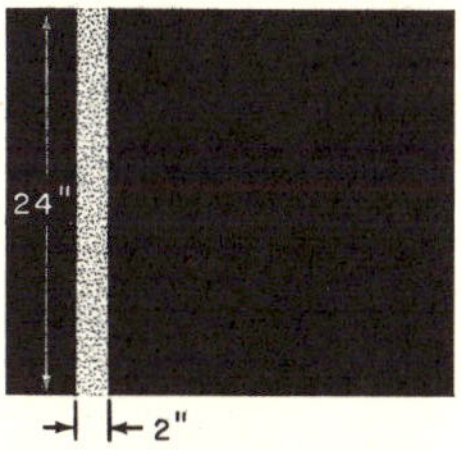

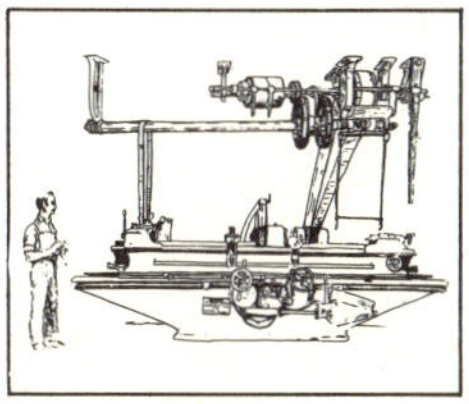
About 1900

0.001 — 1

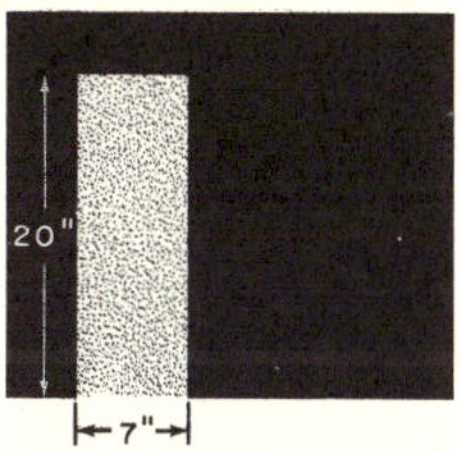

About 1913

0.0005 — 2

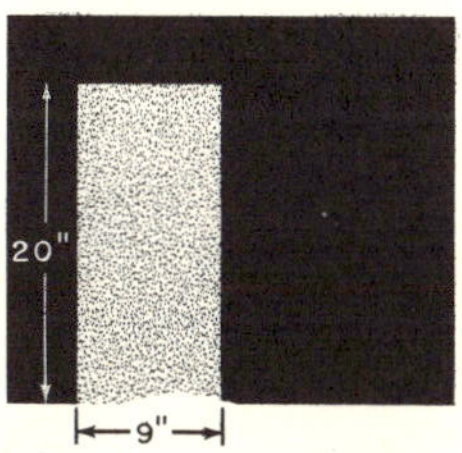

About 1923

0.0005 — 2.5

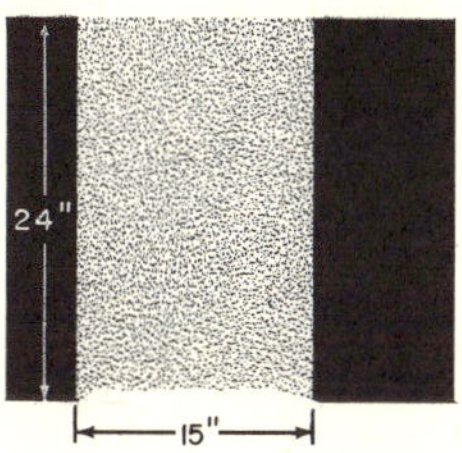

About 1929

0.0003 — 4.5